Corrosion and Materials in the Oil and Gas Industries

Corrosion and Materials in the Oil and Gas Industries

Editor

Sidharth Saini

Corrosion and Materials in the Oil and Gas Industries

Edited by **Sidharth Saini**

Printed in 2017

ISBN: 978-1-68117-037-4

Library of Congress Control Number: 2015931838

© 2016 by

SCITUS Academics LLC,
616, Corporate Way, Suite 2, 4766,
Valley Cottage, NY 10989

www.scitusacademics.com

Contents

Preface

This book gives a comprehensive review of corrosion problems during oil and gas production and its mitigation. The effect of corrosion in the oil industry leads to the failure of parts. This failure results in shutting down the plant to clean the facility. The chemistry of corrosion mechanism had been examined with the various types of corrosion and associated corroding agents in the oil and gas industry. Factors affecting each of the various forms of corrosion were also presented. Ways of mitigating this menace with current technology of low costs had been discussed. It was noticed that the principles of corrosion must be understood in order to effectively select materials and to design, fabricate, and utilize metal structures for the optimum economic life of facilities and safety in oil and gas operations. Also, oil and gas materials last longer when both inhibitors and protective coatings are used together than when only batch inhibition was used. However, it is recommended that consultations with process, operations, materials, and corrosion engineers are necessary in the fitness of things to save billions of dollars wasted on corrosion in the oil and gas industries.

This book covers a wide spectrum of corrosion topics with rich illustrations, an easy to understand writing style, and the presentation of review articles, providing state-of-the-art corrosion-mitigation techniques useful for practicing engineers, technologists, and field technicians.

Editor

Application of Corrosion Inhibitors for Steels in Acidic Media for the Oil and Gas Industry: A Review

Matjaž Finšgar[a] and Jennifer Jackson[b]

[a]University of Maribor, Faculty of Chemistry and Chemical Engineering (UM FKKT), Smetanova ulica 17, 2000 Maribor, Slovenia

[b]BASF SE, EV/ET, Corrosion and Production Chemicals, Global Oilfield Solution, Carl-Bosch Strasse 38, 67056 Ludwigshafen, Germany

ABSTRACT

This review summarizes the corrosion inhibition of steel materials in acidic media. Numerous corrosion inhibitors for steels in acidic solutions are presented. The emphasis is on HCl solutions, lower-grade steels, and elevated temperatures. This review is also devoted to corrosion inhibitor formulation design – mixtures of corrosion inhibitors with (mainly) surfactants, solvents, and intensifiers to improve the effectiveness of individual compounds at elevated temperatures. The information presented in this review is useful for diverse industrial fields, primarily for the well acidizing procedure, and secondly for other applications where corrosion inhibitors for steel materials are needed.

INTRODUCTION

Human demand for fossil fuels is still growing even though alternatives to such energy are currently being sought. Oil and natural gas account for 60% of all global energy demands [1]. It is thus not expected that the conventional method of extracting fossil fuels will disappear within the next few decades. The extraction of geothermal water for use as an energy source is also of paramount importance and its usage is increasing. The methods required to maximize production typically comprise formation stimulation and subsequent well cleaning, both of which can induce a corrosive environment for the steel involved, as it is the main construction material of wells.

Corrosion is worth investigating in oilfield applications, because corrosion problems represent a large portion of the total costs for oil and gas producing companies every year worldwide. Moreover, appropriate corrosion control can help avoid many potential disasters that can cause serious issues including loss of life, negative social impacts, and water resource and environmental pollution. Corrosion in oilfields occurs at all stages from downhole to surface equipment and processing facilities. It appears as leaks in tanks, casings, tubing, pipelines, and other equipment [2], [3] and [4]. Corrosion problems are usually connected with operating problems and equipment maintenance, leading to recurrent partial and even total process shutdown, resulting in severe economic losses [5]. Moreover, Garcia-Arriaga et al. [5] reported that the economic costs linked to the corrosion of natural gas sweetening (CO_2 corrosion) and oil refining plants range between 10% and 30% of the maintenance budget.

In the petroleum industry, general and localized corrosion are the most common types of corrosion occurrences. The other large problem in operating pipe flow lines is internal corrosion [6], mainly due to stress corrosion cracking. Martinez et al. [7] claim that the combination of corrosion and erosion is the main problem in pipe deterioration. Also noted recently is an increase in the occurrence of galvanic corrosion problems associated with the use of different dissimilar materials, which has garnered much attention. Wilhelm [8] reported that the most common situation of coupling dissimilar materials in wells consists of a tubing string made of corrosion-resistant alloy in contact with lower-grade steel casing. Moreover, the metal contacts also cause crevice corrosion in the occluded area between tubing and casing.

The primary focus of this review is to summarize different research relating to corrosion and its inhibition regarding mild, carbon, and low-alloy steel – lower-grade steels – in different acidic solutions encountered in the crude oil and natural gas sector. These materials are used in well construction. In the petroleum industry, one facet of the development of new oil and gas production is the stimulation process. Overall, the stimulation process involves many different aspects, including the acidizing portion utilized to stimulate the carbonate reservoir or for dissolving fines. Typically, highly concentrated acids, between 5 and 28 wt.%, are used which make the environment corrosive to mild, carbon, and low-alloy steels. Hydrochloric, hydrofluoric, acetic, or formic acids are injected into the well during the acidizing stimulation process and cause serious corrosion issues. In the absence of corrosion inhibitors (CIs), the general CR (corrosion rate) can be extremely high (>100 mm/y) and can increase exponentially with increasing temperatures and acid concentrations [9]. Due to the extreme corrosion conditions of this process, developed technology can then be translated to other industries. In particular, this can be relevant for acid pickling, industrial cleaning, and acid descaling, where corrosion conditions are usually milder. This may be a secondary source of information for readers of this review. It has to be pointed out that the petroleum industry is the largest consumer of CIs. This review only addresses individual CIs for application in HCl mediums with different steels because HCl is the most prevalent acid used in stimulation.

An effort has been made herein to combine different works by the same authors in a single paragraph, even though not all authors of different articles or patents appear together all the time. In this review, when steel materials in general are written about, lower-grade steels are being referred to. All concentrations in % are always reported as a mass fraction if not stated otherwise. Moreover, when concentrations in various articles were reported in parts per million (ppm), herein they are converted to mg/L.

This work discusses the well acidizing procedure in general so that readers of this review can gain an impression of the severe corrosion conditions during that process. Moreover, the steel materials used for well construction and associated with corrosion problems are discussed. The corrosion of these steel materials and previously tested CIs for HCl solutions are reviewed. This review also explains aspects of a corrosion inhibitor formulation design in order to increase the success of these CIs at elevated temperatures or under other well environmental conditions. Furthermore, it also presents environmental concerns in corrosion inhibition processes, environmental friendly methanesulphonic acid, and some recommendations for correct test methods regarding acid CIs.

THE WELL ACIDIZING PROCEDURE

Limestone formations or carbonate-bearing sandstone carry many hydrocarbon reservoirs [10]. A very important step in the oil, gas, and geothermal water drilling industry is the well acidizing procedure, which is a rock reservoir (the origin of the natural resource or water – a geological subterranean formation) stimulation technique used to improve productivity. Acids are forced under high pressure through the borehole into the pore spaces of the rock formation, where they react chemically with rocks to dissolve them (usually calcite, limestone, and dolomite), which enlarges the existing flow channels and opens new ones to the wellbore [11], [12], [13], [14] and [15]. Acidizing is used in conjunction with hydraulic fracturing techniques and matrix acidizing techniques [16]. In fracture acidizing treatments, one or more fractures are produced in the formation and acidic solution is introduced into the fracture to etch flow channels in the fracture face. The acid also enlarges the pore spaces in the fracture face and in the formation [12] and [13]. The fractures are then filled with sand or other material in order to prevent the fractures from closing and allow the penetration of natural resources or water. Acids are often also employed for scale removal treatments (pickling of the well tubing) and for the removal of drilling mud damage in newly drilled wells before being brought into production [17]. For example, the combination of fluorosilicate with metal ions such as Na^+ may cause the precipitation of gelatinous compounds, which need to be removed [10]. Scale removal treatments are usually done with 15% HCl at temperatures up to 60 °C in order to remove iron oxides and carbonated minerals [18]. Acidizing steps are frequently repeated. All these procedures involve the injection of acids into the well system made of steel tubes. In deep wells the downhole temperature may exceed 200 °C [19] and [20]. During the acidizing process metallic materials can also come into contact with acid solution and sometimes with H_2S and CO_2 at elevated temperatures. Due to the above listed problems, the acidizing process requires a high degree of corrosion protection of tubular materials and other equipment employed.

The Use of Acids in the Acidizing Procedure

Different acids are employed depending on the underground reservoir characteristics. The treatment normally involves the injection of acid at 15% concentration (sometimes from 5% up to 28%) [11], [21], [22] and [23]. A standard 15% acid concentration had been chosen before 1960 due to the insolubility of arsenic inhibitor, the primary inhibitor of the industry at the time, because it was not soluble in HCl concentrations higher than 17% [24]. The most common conventional acids are HCl, HF, acetic, and formic acids. It has also been noted that mixtures of these conventional acids with sulphamic, sulphuric, phosphoric, methanesulphonic, nitric, citric, and chloroacetic acids are employed [9], [12], [13], [15], [25], [26], [27] and [28].

The majority of acidizing treatments carried out utilize HCl at concentrations of 5–28% [24]. HCl has an advantage over the other mineral acids in the acidizing operation because it forms metal chlorides, which are very soluble in the aqueous phase. Other acids have been employed historically, but they were not so successful compared with HCl. One of the reasons was that, for example, sulphate, nitrate, and phosphate salts have lower solubility compared with chloride salts in aqueous media [29]. HCl is widely used for stimulating carbonate-based reservoirs such as limestone and dolomite. Alternatively, sandstone formation can occur and for a successful stimulation process HF is needed. Sometimes a combination of HCl with HF, also called mud acid, is employed (typically 6% HCl/1% HF and 12% HCl/3% HF) [20] and [23]. For such application three treatment procedures are typically used: a pre-flush, main-flush, and after-flush. The pre-flush is done with 15% HCl. The main-flush is done with 12% HCl and 1.5% HF. The after-flush is performed for rapid formation clean-up. This can be done using ethylene glycol monobutyl ether (EGMBE) or methanol.

Moreover, HCl represents the most economical acid for dissolving $CaCO_3$ in pickling applications. However, the fast reaction rate with rocks, the CR, and the pitting tendency of materials vary considerably with HCl concentration, which can cause problems. Additionally, another disadvantage of using HCl is its high corrosiveness to steel, aluminium, or chromium-plated equipment.

One important factor that is considered in acid stimulation is the reaction rate of the acid in the formation. This rate is dependent on the type of acid, acid concentration, temperature, fluid velocity, and the type of formation material. Acidic solutions are introduced into the formation and can only travel a short distance before they become spent. The acid pumped down the well is usually called the live acid. The acid that is produced from the well after completion of the treatment is usually called spent acid. Due to the reaction of the acid with rocks, the pH of the spent acid goes up, commonly reaching a value near 1. An example of a spent acid simulation in the laboratory is the use of HCl with 150 g of dissolved $CaCl_2$ and a pH adjusted to 1[10].

It is desirable to maintain the acidic solution in a reactive state for as long as possible to maximize the permeability enhancement [12], [13] and [28]. On the other hand, in order to achieve the desired deeper penetration of the stimulation fluid into the rock formation, the acid needs to be emulsified by an appropriate agent. In such a manner the acid reaction rate (especially HCl) with the rocks is significantly retarded, but still effective, and the acid is spent more slowly, allowing deeper penetration ("unretarded" 15% HCl at ~160 °C would penetrate only approximately 10 cm) [19]. In some applications formic and acetic acids can be used in conjunction with HCl since they react more slowly compared with HCl by itself. For example, combinations of 9% formic acid and 13% acetic acid with 15% HCl are employed [19]. This combination of acids helps to improve the fracture geometry. However, Ali et al. [15] reported that formic acid and other short-chain aliphatic acids and their related aldehydes are still as corrosive as HCl to pipelines and/or other equipment. Fortunately, they are easier to inhibit compared with HCl by itself [19]. Unfortunately, the solubility of carbonates in these acids is much lower than in HCl. Formic and acetic acids may be used instead of HCl in limited applications at very high temperature [24]. In particular, this is relevant in wells completed with Super chrome-13 tubulars [30]. At higher formic and acetic acid concentrations, potential precipitation of reaction products of the acid with the rock formation is expected (mainly calcium formate or acetate) [30]. Consequently, this will lead to deposit problems and disturb or even prevent the flow of fluids.

In summary, the corrosion of pipelines and other equipment involved in the oil and gas industry represent a large problem in the acidizing process and consequently a large part of the total cost and potential danger to the personnel involved. Thus, the selection of non-corrosive or low-corrosive inhibited acid solution is crucial.

MATERIALS USED FOR WELL CONSTRUCTION AND IN CORROSION TESTS

The materials used for pipeline construction play an important role in the petroleum industry as they carry liquids and gases over long distances from their source to the ultimate consumers. Corrosion problems associated with the transportation process can exist at every stage of production, from the initial extraction to refining and storage prior to use.

The steels used in well construction can range from mild steel API N80 (API – American Petroleum Institute), L80, and J55, to high Cr-content corrosion resistant alloys, such as austenitic-ferritic steel, e.g. duplex 22% Cr steel, and modified martensitic 13% Cr steel, sometimes called Super-13 [18] and [20]. Moreover, an important property of pipe corrosion resistance is how the weld of the pipe is made and the acid corrosion at that place [1] and [31].

Over the past few decades there have been many developments in new corrosion resistant alloys (CRA), however CSs (carbon steels) are still the most commonly used materials for downhome tubular, flow lines, and transmission pipelines in the gas and oil industry, most likely due to their low cost [4], [6], [32] and [33]. For example, the cost of austenitic stainless steels (AISI 304 and 316) is currently around 8-times higher than that of CSs [1]. A combination of CS and chemical treatments is the most cost-effective method for corrosion control [34]. Without the usage of chemical treatments, in particular CIs, CS materials are highly susceptible to corrosion in most acids [35]. More corrosion resistant alloys may also be employed, e.g. austenitic or duplex stainless steels [23]. However, high-grade alloys significantly increase the capital cost and are susceptible to corrosion in media containing large amounts of chloride ions [18] and [34].

API N80 CS has generally been used as the main construction material for downhole tubulars, flow lines, and transmission pipelines in the petroleum industry [13], [17], [36] and [37] and consequently most acid CI data exist for that steel type [24]. API L-80 grade CS tubing is H_2S-resistant steel [23]. Moreover, Torres-Islas et al. [38] reported that micro-alloyed API X80 steel was specially designed for sour gas transport and is intended to be applied as pipeline steel in Mexico. API P-110 was reported to be used as the production liner, whereas API G95 steel has been used for the tubing [19].

Steel metallurgy is one of the most important criteria for acidizing CIs in laboratory testing when simulating real field conditions. A problem sometimes arises because the standard materials commonly employed, e.g. API N80 steel, may vary considerably from one manufacturer to another or from one lot to the next, which leads to confusion in comparative corrosion testing. In order to clearly show the differences in steel composition, even though these steels are sometimes named the same, Table 1 was constructed for the materials found in the literature (presented below). Primarily for steel producers, tubing must meet tensile stress requirements, and that is why the CR of steel types with the same name can vary considerably. Smith et al. [24] claim that the variance in the CR of different steel materials (most likely they meant carbon or low-alloy) can be up to 35% at temperatures above 93.3 °C.

Table 1: Chemical compositions in wt.% of different iron-based materials

	Designation API/EN/AISI/UNS/ etc.	C	Mn	Cr	Mo	Si	P	S	Ni	V	Nb	Cu	Al	Ti	Pb	W	Co	As	N	B	Sn	Fe	References
CS	API N80	0.31	0.92	0.20	–	0.19	0.01	0.008	–	–	–	–	–	–	–	–	–	–	–	–	–	Balance	[127]
CS	API N80	0.42	1.55	0.051	0.18	0.24	0.012	0.004	0.005	–	–	0.06	–	0.01	–	–	–	–	–	–	–	Balance	[37]
–	API N80	0.52	1.50	–	–	0.23	0.011	0.008	0.07	–	<0.005	0.07	0.01	–	–	–	–	–	–	–	–	Balance	[3]
–	API N80	0.24	1.19	0.036	0.021	0.22	0.013	0.004	0.028	0.017	0.006	0.019	–	0.011	–	–	–	–	–	–	–	Balance	[144]
–	N80	0.31	0.92	0.20	–	0.19	0.01	0.008	–	–	–	–	–	–	–	–	–	–	–	–	–	Balance	[36]
–	N-80	0.028	1.48	0.2	0.1	0.17	0.015	0.015, 0.026*	–	–	–	–	0.007	–	–	–	–	–	–	–	–	Balance	[65]
CS	N80	0.23	1.35	0.05	0.01	0.22	0.017	0.012	0.07	–	–	–	–	–	–	–	–	–	–	–	–	Balance	[32]
CS	N80	0.31	0.92	0.20	–	0.19	0.01	0.008	–	–	–	–	–	–	–	–	–	–	–	–	–	Balance	[17]
CS	API 180	0.43	1.90	0.16	–	0.35	–	0.007	0.20	–	–	–	–	–	–	–	–	–	–	–	–	Balance	[145]
–	API X52	0.07	1.27	0.017	0.013	0.27	0.015	0.007	0.023	–	0.055	0.053	0.036	0.002	–	–	–	–	–	–	–	Balance	[7]
–	API X65	0.07	1.35	0.16	0.17	0.24	0.017	0.005	0.18	–	–	0.01	–	–	–	–	–	–	–	–	–	Balance	[146]
–	API X65	0.04	1.5	–	0.07	0.2	0.011	0.003	–	–	–	–	–	–	–	–	–	–	–	–	–	Balance	[35]
–	API X80	0.044	1.69	0.0103	0.25	0.271	0.0091	0.0003	0.240	0.0010	0.0554	0.1170	0.0310	0.014	–	–	–	–	–	0.0001	–	Balance	[38]
–	P110	0.26	1.40	0.15	0.01	0.20	0.009	0.003	0.012	0.012	0.006	<0.01	–	0.03	–	–	–	–	–	–	–	Balance	[144]
–	J55	0.19	1.39	0.19	0.097	0.31	0.014	0.004	0.017	0.013	0.007	<0.01	–	0.04	–	–	–	–	–	–	–	Balance	[144]
–	API J55	0.41	1.11	0.03	0.03	0.05	0.008	0.018	0.23	–	–	0.03	–	–	–	–	–	–	–	–	–	Balance	[108]
–	API J55	0.41	1.11	0.05	0.05	0.05	0.0075	0.018	0.02	–	–	0.02	–	–	–	–	–	–	–	–	–	Balance	[109]
–	API J55	0.29	1.52	0.04	0.13	0.08	0.013	0.071	<0.02	–	–	<0.02	–	–	–	–	–	–	–	–	–	Balance	[60]
CS	UNS-G4130	0.270	0.82	0.88	0.44	0.25	0.012	<0.01	0.014	–	–	0.18	–	–	–	–	–	–	–	–	–	Balance	[18]
CS	DIN C835	0.348	0.63	0.08	0.017	0.35	0.009	0.007	0.09	–	–	–	0.01	–	–	–	–	–	–	–	–	Balance	[147]
CS	C-1018	0.21	0.05	–	–	0.38	0.09	0.05	–	–	–	–	–	–	–	–	–	–	–	–	–	Balance	[148]
LCS	HS80	0.1–0.15	0.6–0.9	0.45–0.70	–	0.1–0.5	<0.03	<0.005	<0.25	–	–	<0.4	0.01	–	–	–	–	–	–	–	–	Balance	[9]
LCS	HS110	0.1–0.15	0.6–0.9	0.55–0.70	0.25–0.45	0.25–0.40	<0.025	<0.005	0.14–0.30	–	–	<0.4	–	–	–	–	–	–	–	–	–	Balance	[9]
LCS	DIN EN 10130:99	0.07	0.35	–	–	0.1	0.015	0.015	–	–	–	–	–	–	–	–	–	–	–	–	–	Balance	[89]
LCS	X-65	0.065	1.54	0.05	0.007	0.25	0.013	0.001	0.04	–	–	0.04	0.041	–	–	–	–	–	–	–	–	Balance	[149]
LCS	St52	0.130	1.25	0.12	0.02	0.35	0.027	0.004	0.08	–	–	0.31	0.035	–	–	–	–	–	0.003	–	–	Balance	[149]
CS	1018	0.18	0.71	–	–	0.19	0.009	–	–	–	–	–	–	–	–	–	–	–	–	–	–	Balance	[39]
CS	UNS G10180	0.18	0.72	–	–	0.10	0.007	0.01	–	–	–	–	–	–	–	–	–	–	–	–	–	Balance	[143]
MS	–	0.16	0.40	–	0.007	–	0.013	0.07	–	–	–	–	–	–	–	–	–	–	–	–	–	Balance	[90]
MS	–	0.089	0.34	0.037	–	–	0.01	0.079	0.027	0.005	–	0.005	–	–	–	–	–	–	–	–	–	99.47*	[91]
MS	–	0.184	0.29	0.097	0.071	0.07	0.017	0.014	0.071	0.014	–	0.065	–	–	–	–	–	–	–	–	–	99.15	[91]
MS	–	0.18	0.6	–	–	0.1	0.04	0.05	–	–	–	–	–	–	–	–	–	–	–	–	–	Balance	[98]
MS	–	0.049	0.277	2.34	–	–	–	0.0005	–	–	–	–	–	–	–	–	–	–	–	–	–	Balance	[150]
MS	–	0.179	0.439	–	–	0.166	–	0.014	–	–	–	0.203	–	–	–	–	–	–	–	–	–	Balance	[151]

Material			C																									Fe	Ref.
MS	–	0.37	–	0.078	0.002	0.04	0.073	0.02	0.017																			Balance	[102]
MS	–	0.380	–	0.770		0.080	0.008	0.006						0.53	0.07													99.2%	[29]
MS	–	0.554	0.04	0.077	0.03	0.666	0.018					0.06	0.019	0.06														Balance	[75]
MS	–	0.46	–	0.01		0.35	0.018	–	0.017																			Balance	[151]
MS	–	0.35	0.06	0.17	–	0.26	0.01	0.017			0.07	0.10																Balance	[152 and 153]
MS	–	0.35	0.06	0.16	–	0.016	0.03	0.079	–		0.019	0.10																Balance	[73]
MS	–	0.35	–	0.16	–	0.06	0.05	0.079			0.10																	Balance	[74]
MS	–	0.39	–	0.17	–	0.09	0.05	0.055			0.10																	Balance	[96]
MS	–	0.05	–	0.13	–	0.04	0.04				0.025																	0.9?	[81]
MS	–	0.39	–	0.21	–	0.38	0.09	0.05				0.01																Balance	[97]
MS	–	0.29	–	0.13	–	0.18	0.005	0.04			0.025																	Balance	[93], [93], [94],[95] and [96]
CS	–	0.60–0.90	–	0.2	–	0.15–0.30	0.04	0.05																				Balance	[154]
CS	–	0.27–0.63	–	0.06–0.18	–		0.04–0.05	0.4–0.05																				Balance	[80]
CS	–	0.35	–	0.18	–	Max 0.045	Max 0.045	Max 0.058																				Balance	[4]
CS	–	0.150	–	0.18	–	0.17	0.03	0.0025																				Balance	[155]
CS	–	0.5	0.06	0.200	–	0.003	0.074	0.06												0.2								Balance	[79]
CS	–	0.35	0.06	0.2	–	0.15	0.04	0.025																				Balance	[33]
CS	–	0.7	0.015	0.18	–	0.17	0.005	0.001												0.017		0.002						98.36	[156]
CS	–	0.35	–	0.06	–	0.06	0.005	0.025			0.02																	Balance	[84]
CS	–	0.50	–	0.14	–	0.17	0.025	0.025																				Balance	[114]
CS	–	0.7	–	0.10	–	0.005	0.025	0.025																				Balance	[99]
CRP	–	0.35	–	0.14	0.004	0.17	0.03	0.008						0.003														Balance	[83]
CRP	–							0.05																				Balance	[157]
CRMS	–	0.799	–	0.12	–	0.17	0.03	0.025			0.025			0.004														Balance	[11], [271], [162], [65], [67], [87] and [88]
CRMS	–	0.5	0.014	0.003	0.001	0.102	0.033	0.018	0.019	0.001	0.011	0.004	0.007	0.002	0.006	0.001	0.006											99.5	[134], [103]

INHIBITION OF STEEL CORROSION

The Use of Corrosion Inhibitors in Acidizing Treatments

It is almost impossible to prevent corrosion, however it is possible to control it [23]. To control the corrosion damage of well tubulars, mixing tanks, coiled tubings, and other metallic surfaces, acids need to be inhibited by the use of an effective CI solution (now commonly organic compounds) [18] and [21].

A CI is a chemical substance that is effective in very small amounts when added to a corrosive environment to decrease the CR of the exposed metallic material. Unfortunately, CIs are effective only for a particular metallic material in a certain environment. Minor changes in the composition of the solution or alloy can significantly change the inhibition effectiveness ( ). For example, many good inhibitors that in the past worked for 15% HCl did not perform well for 28–30% HCl. On the other hand, Smith et al. [24] also reported that some inhibitors developed for concentrated acid are not as effective in less concentrated HCl. However, this should rarely occur.

For stimulation applications, inhibitors are added to the acid fluids in batch-wise fashion; batch-wise refers to the single addition of the CI into the holding tank of the acid before the acid is used in the stimulation process. The use of CIs is one of the most cost-effective and practical methods of corrosion protection. The employment of an appropriate inhibitor can allow the use of lower-grade CSs, which significantly reduces the capital costs of a well construction project when compared with the use of high-grade alloys in the same project [39]. The selection and the amount of CI used depend on the acid type and its strength, the steel type, the desired protection time, and the expected temperature [23]. The maximum temperature limit is one of the key roles in CI selection, because some components are sensitive to thermal decomposition, i.e. when they lose their inhibition effectiveness [10]. Smith et al. [24] claim that the introduction of arsenic acid as a CI in 1932 was responsible for the development of well acidizing. However, for arsenic compounds it is known that they produce poisonous arsine gas under acidic conditions [40] and numerous persons died in the past due to arsenic poisoning [41]. Subsequently, the majority of CIs used were inorganic salts or acids such as arsenate or arsenic acid until the mid-1970s, when they were replaced by organic molecules, which generally contain N, O, P, or S heteroatoms, of an aromatic and/or unsaturated character [9], [42],[43], [44], [45], [46], [47], [48], [49], [50], [51], [52], [53], [54], [55], [56], [57] and [58].

The corrosion environments in oil and gas production wells can be highly variable, which makes the selection and application of inhibitors complicated. Frequently it happens that a CI that works in one well may not work in another well [59]. Numerous compounds used as CIs are discussed below. Because the mechanism of how CIs work is usually not known, empirical testing is still unavoidable, in spite of some proposed models for forecasting . The scientific community and the industry do not fully understand the mechanism or the role of CIs and it is difficult or sometimes impossible to predict if a particular compound will work or not [55]. Walker [12], [13] and [28] claims that in general CIs are effective for ferrous materials only at temperature levels below 121–149 °C.

Corrosion Inhibitor Compounds

A variety of organic compounds act as CIs for steels during the acidizing procedure, including acetylenic alcohols, aromatic aldehydes, alkenylphenones [25], [27], [60], [61] and [62], amines [13], amides, nitrogen-containing heterocycles (e.g. imidazoline-based [3], [35] and [63]), nitriles, imminium salts, triazoles, pyridine and its derivatives or salts [14], [16] and [64], quinoline derivatives, thiourea derivatives, thiosemicarbazide, thiocyanates, quaternary salts [14], [16] and [64], and condensation products of carbonyls and amines ([11], [18], [65], [66] and [67] and the refs. therein). Molecules containing nitrogen and acetylenic alcohols are claimed to form a film on the metal surface and can retard the metal dissolution process (an anodic reaction) as well as hydrogen evolution (a cathodic reaction) [9]. However, it has been reported that propargyl alcohol is soluble in acids, but the solubility of other acetylenic alcohols decreases with increasing carbon chain length. On the other hand, the solubility of such acetylenic alcohols can be increased when combined with quaternary ammonium surfactants [9] (explained below). Acetylenic alcohols are widely used because of their commercial availability and cost effectiveness. Propargyl alcohol is usually taken as a standard CI for acidization [65] and sometimes it has a significant synergistic effect with other compounds. In our experience, the most commonly used CIs in the natural resource exploitation industry are propargyl alcohol and its derivatives, cinnamaldehyde, and nitrogen aromatic-based compounds such as pyridinium benzyl quaternary chloride.

In 1984, a comprehensive review article was published by Schmitt [66] which presented the application of CIs for acid media. Due to this reason, the focus of this review is the literature published subsequently, most of which is summarized in Table 2, where the investigated corrosive medium, steel types, pH (if reported), the concentration of CI used, the method of corrosion testing, and the minimum and maximum reported or CR are summarized (because the different experimental techniques used sometimes resulted in quite large differences). This review includes corrosion studies on HCl mostly related to the petroleum industry for low Cr content steels (e.g. MS – mild steel – and CS). Moreover, only studies published for HCl solution are reviewed below, because the majority of acidizing jobs are performed by this acid.

Table 2: The inhibition effectiveness, , of different CIs, or the CR of different steel materials in various solutions. The values in this table are reported only if quoted (as numbers) in the text of the article (no calculation of these values were made). In case different techniques for or CR determination were used (frequently different values were obtained for the same test conditions), this table reports the range from the minimum to the maximum values reported in the article. The concentration in % is always related to the mass fraction if not stated otherwise. LPR and EFM stand for linear polarization resistance and electrochemical frequency modulation, respectively

Inhibitor	Medium	Material	Inhibitor concentration	Reported pH, special treatment, test conditions, testing technique	(%) (or CR)	References
Benzoin	0.1–0.5 mol/L HCl	MS	0.1–0.5 mmol/L	WL at 30 and 40 °C	56–89%	[68][a]
Benzoin	8 mol/L HCl	MS	0.5 mmol/L	Measurement of H$_2$ gas evolution	47%	[68][a]
Benzoin-(4-phenylthiosemicarbazone)	0.1–0.5 mol/L HCl	MS	0.1–0.5 mmol/L	WL at 30 and 40 °C	18–60%	[68][a]

Benzoin-(4-phenylthiosemicarbazone)	8 mol/L HCl	MS	0.5 mmol/L	Measurement of H_2 gas evolution	34.5%	[68][a]
Benzyl	0.1–0.5 mol/L HCl	MS	0.1–0.5 mmol/L	WL at 30 and 40 °C	20–59%	[68][a]
Benzyl	8 mol/L HCl	MS	0.5 mmol/L	Measurement of H_2 gas evolution	24.5%	[68][a]
Benzyl-(4-phenylthiosemicarbazone)	0.1–0.5 mol/L HCl	MS	0.1–0.5 mmol/L	WL at 30 and 40 °C	12–55%	[68][a]
Benzyl-(4-phenylthiosemicarbazone)	8 mol/L HCl	MS	0.5 mmol/L	Measurement of H_2 gas evolution	19%	[68][a]
-Pyridoin	0.5 mol/L HCl	MS	0.01–0.5 mmol/L	WL at 30 and 40 °C	CR = 8.6–55.7 mg/dm² day	[85]
2,2'-Pyridil	0.5 mol/L HCl	MS	0.01–0.5 mmol/L	WL at 30 and 40 °C	CR = 17.9–71.4 mg/dm² day	[85]
Sodium N-1-n-hexyl-phthalamate	0.5 mol/L HCl	CS SAE 1018	0.037–0.339 mmol/L	WL at T = 25–40 °C	15–80%	[69][a]
Sodium N-1-n-hexyl-phthalamate	0.5 mol/L HCl	CS SAE 1018	0.037–0.339 mmol/L	Tafel extrapolation at room T	22–69%	[69][a]
Sodium N-1-n-decyl-phthalamate	0.5 mol/L HCl	CS SAE 1018	0.030–0.306 mmol/L	WL at T = 25–40 °C	20–83%	[69][a]
Sodium N-1-n-decyl-phthalamate	0.5 mol/L HCl	CS SAE 1018	0.030–0.306 mmol/L	Tafel extrapolation at room T	29–63%	[69][a]
Sodium N-1-n-tetradecyl-phthalamate	0.5 mol/L HCl	CS SAE 1018	0.026–0.261 mmol/L	WL at T = 25–40 °C	25–86%	[69][a]
Sodium N-1-n-tetradecyl-phthalamate	0.5 mol/L HCl	CS SAE 1018	0.026–0.261 mmol/L	Tafel extrapolation at room T	23–65%	[69][a]
Benzimidazole	Deaerated 1 mol/L HCl	MS	20 mmol/L	Tafel extrapolation T = 20–60 °C	29.5–60%	[72]
2-Aminobenzimidazole	Deaerated 1 mol/L HCl	MS	20 mmol/L	Tafel extrapolation T = 20–60 °C	84.0–86.8%	[72]
2-Mercaptobenzimidazole	Deaerated 1 mol/L HCl	MS	1 mmol/L	Tafel extrapolation T = 20–60 °C	93.9–97.0%	[72]
1-Benzylbenzimidazol	Deaerated 1 mol/L HCl	MS	5 mmol/L	Tafel extrapolation T = 20–60 °C	97.2–97.8%	[72]
1,2-Dibenzylben-zimidazole	Deaerated 1 mol/L HCl	MS	1 mmol/L	Tafel extrapolation T = 20–60 °C	97.4–98.2%	[72]
Indole	1 mol/L HCl	MS[b]	2 mmol/L	WL and LSW	92.5%	[74]
1H-benzotriazole	1 mol/L HCl	MS[b]	2 mmol/L	WL and LSW	87.1%	[74]
1,3-Benzothiazole	1 mol/L HCl	MS[b]	2 mmol/L	WL and LSW	85.7%	[74]
Benzimidazole	1 mol/L HCl	MS[b]	2 mmol/L	WL and LSW	60.0%	[74]
Mercapto-triazoles (4 compounds)	1 mol/L HCl	MS[b]	0.4 g/L	Tafel extrapolation, EIS and WL at 25 °C	91.4–98.7%	[90][a]

Benzaldehyde	1 mol/L HCl	MS[b]	0.4 g/L	Tafel extrapolation at 25 °C	61.4%	[90][a]
4-Amino-5-mercapto-1,2,4-triazole	1 mol/L HCl	MS[b]	0.4 g/L	Tafel extrapolation at 25 °C	68.4%	[90][a]
3-Phenyl-4-amino-5-mercapto-1,2,4-triazole	1 mol/L HCl	MS[b]	0.4 g/L	Tafel extrapolation at 25 °C	88.8%	[90][a]
Different isoxazolidines	1 mol/L HCl	MS[b]	50–400 mg/L	WL and Tafel extrapolation at 60 °C	40.7–99.5%	[91]
1,12-Bis(1,2,4-triazolyl) dodecane	Deaerated and aerated 1 mol/L HCl	CS[b]	0.01–10 mmol/L	Tafel extrapolation and EIS	55–93%	[80]
Octyl alcohol	15% HCl	MS[b]	0.2–1%	Tafel extrapolation, EIS and WL at 35 and 105 °C	48–87%	[29]
Propargyl alcohol	15% HCl	MS[b]	0.2–1%	Tafel extrapolation, EIS and WL at 35 and 105 °C	97–100%	[29]
Benzimidazole	1 mol/L HCl	MS[b]	50–250 mg/L	Tafel extrapolation and EIS at 25 °C	36.6–73.8%	[75]
Benzimidazole	1 mol/L HCl	MS[b]	250 mg/L	Tafel extrapolation and EIS at 25–55 °C	37.3–52.2%	[75]
2-Methylbenzimidazole	1 mol/L HCl	MS[b]	50–250 mg/L	Tafel extrapolation and EIS at 25 °C	43.9–76.3%	[75]
2-Methylbenzimidazole	1 mol/L HCl	MS[b]	250 mg/L	Tafel extrapolation and EIS at 25–55 °C	40.0–57.1%	[75]
2-Mercaptobenzimidazole	1 mol/L HCl	MS[b]	50–250 mg/L	Tafel extrapolation and EIS at 25 °C	72.6–90.4%	[75]
2-Mercaptobenzimidazole	1 mol/L HCl	MS[b]	250 mg/L	Tafel extrapolation and EIS at 25–55 °C	74.7–88.8%	[75]
4-Methylpiperidine	Aerated 1 mol/L HCl	Pure Fe	0.1–10 mmol/L	Tafel extrapolation and EIS at 25 °C	54.31–87.52%	[78]
4-Benzylpiperidine	Aerated 1 mol/L HCl	Pure Fe	0.1–10 mmol/L	Tafel extrapolation and EIS at 25 °C	53.55–83.82%	[78]
Piperidine	Aerated 1 mol/L HCl	Pure Fe	0.1–10 mmol/L	Tafel extrapolation and EIS at 25 °C	49.37–75.58%	[78]
3-Methylpiperidine	Aerated 1 mol/L HCl	Pure Fe	0.1–10 mmol/L	Tafel extrapolation and EIS at 25 °C	37.19–70.42%	[78]

2-Methylpiperidine	Aerated 1 mol/L HCl	Pure Fe	0.1–10 mmol/L	Tafel extrapolation and EIS at 25 °C	34.78–63.39%	[78]
3,5-dimethylpiperidine	Aerated 1 mol/L HCl	Pure Fe	0.1–10 mmol/L	Tafel extrapolation and EIS at 25 °C	46.98–54.86%	[78]
Cis-2,6-dimethylpiperidine	Aerated 1 mol/L HCl	Pure Fe	0.1–10 mmol/L	Tafel extrapolation and EIS at 25 °C	33.12–53.91%	[78]
2-Chloroaniline	Aerated 1 mol/L HCl	Pure Fe	1–10 mmol/L	Tafel extrapolation and EIS at 25 °C	63.62–80.64%	[77]
2-Fluoroaniline	Aerated 1 mol/L HCl	Pure Fe	1–10 mmol/L	Tafel extrapolation and EIS at 25 °C	55.04–71.79%	[77]
2-Methoxyaniline	Aerated 1 mol/L HCl	Pure Fe	1–10 mmol/L	Tafel extrapolation and EIS at 25 °C	54.67–66.83%	[77]
2-Ethylaniline	Aerated 1 mol/L HCl	Pure Fe	1–10 mmol/L	Tafel extrapolation and EIS at 25 °C	53.37–63.24%	[77]
2-Ethoxyaniline	Aerated 1 mol/L HCl	Pure Fe	1–10 mmol/L	Tafel extrapolation and EIS at 25 °C	49.88–64.25%	[77]
2-Methylaniline	Aerated 1 mol/L HCl	Pure Fe	1–10 mmol/L	Tafel extrapolation and EIS at 25 °C	43.13–60.09%	[77]
2-Aminoben-zimidazole	1 mol/L HCl	Pure Fe	1–10 mmol/L	Tafel extrapolation and EIS at 25 °C	63.38–78.28%	[76]
2-(2-Pyridyl) benzimidazole	1 mol/L HCl	Pure Fe	1–10 mmol/L	Tafel extrapolation and EIS at 25 °C	58.65–72.40%	[76]
2-Aminomethy-lbenzimidazole	1 mol/L HCl	Pure Fe	1–10 mmol/L	Tafel extrapolation and EIS at 25 °C	56.28–68.24%	[76]
2-Hydroxyben-zimidazole	1 mol/L HCl	Pure Fe	1–10 mmol/L	Tafel extrapolation and EIS at 25 °C	45.50–58.05%	[76]
Benzimidazole	1 mol/L HCl	Pure Fe	1–10 mmol/L	Tafel extrapolation and EIS at 25 °C	41.33–51.07%	[76]
Different acetamide derivatives	2 mol/L HCl (+10% acetone)	Cold rolled LCS DIN EN 10130–99[b]	35–100 mg/L	WL at room T	58.3–91.2%	[89]
Different isoxazolidine derivatives	2 mol/L HCl (+10% acetone)	cold rolled LCS DIN EN 10130–99[b]	35–100 mg/L	WL at room T	46.9–91.0%	[89]

Different isoxazoline derivatives	2 mol/L HCl (+10% acetone)	cold rolled LCS DIN EN 10130–99[b]	35–150 mg/L	WL at room T	26.9–91.6%	[89]
2-Aminomethy-lbenzimidazole	Deaerated 0.5 mol/L HCl	CS[b]	0.2–2 mmol/L	LSW and EIS	69.0–90.0	[79]
Bis(benzimidazol-2-ylethyl) sulphide	Deaerated 0.5 mol/L HCl	CS[b]	1×10^{-4} to 1 mmol/L	LSW and EIS	16.0–98.0%	[79]
3-(4-Amino-2-methyl-5-pyrimidyl methyl)-4-methyl thiazolium chloride	0.5 mol/L HCl	MS[b]	0.01–1 mmol/L	WL at 30 °C	35.7–78.0%, CR = 0.4–1.2 mm/y	[81]
Glycine	1 mol/L HCl	Cold rolled steel[b]	0.1–5 mmol/L	LPR, Tafel extrapolation, EIS and EFM at 25 °C	11.0–75.0%, CR = 0.26–1.08 mm/y	[158]
2-(Bis(2-aminoethyl)amino) acetic acid	1 mol/L HCl	Cold rolled steel[b]	0.1–5 mmol/L	LPR, Tafel extrapolation, EIS and EFM at 25 °C	16.5–97.2%, CR = 0.03–1.08 mm/y	[158]
Cinnamaldehyde	20% HCl	N80 steel	1%	WL at 90 °C	CR = 6.3 g/m² h	[124]
Cinnamaldehyde (containing 10% propargyl alcohol)	20% HCl	N80 steel	1%	WL at 90 °C	CR = 2.9 g/m² h	[124]
Benzalacetone	20% HCl	N80 steel	1%	WL at 90 °C	CR = 116.0 g/m² h	[124]
Benzalacetone (containing 10% propargyl alcohol)	20% HCl	N80 steel	1%	WL at 90 °C	CR = 6.9 g/m² h	[124]
Chalcone	20% HCl	N80 steel	1%	WL at 90 °C	CR = 412.5 g/m² h	[124]
Chalcone (containing 10% propargyl alcohol)	20% HCl	N80 steel	1%	WL at 90 °C	CR = 14.7 g/m² h	[124]
Different 3,5-bis(n-pyridyl)-4-amino-1,2,4-triazoles (n = 1, 2, 3)	1 mol/L HCl	MS[b]	100–500 mg/L	WL and EIS at 30 °C	75.9–98.8%	[92]
1,4-Bis(2-pyridyl)-5H-pyridazino[4,5-b]indole	1 mol/L HCl	MS[b]	0.01–0.1 mmol/L	WL, LSW and EIS at 30 °C	50.5–94.0% CR = 0.36–2.09 mg/cm² h	[94]
3,5-Bis(2-thienyl)-1,3,4-thiadiazole	Aerated and deaerated 1 mol/L HCl	MS[b]	0.025–0.15 mmol/L	WL, LSW and EIS at 30–60 °C	80.3–98.2% CR = 0.32–0.90 mg/cm² h	[95] and [96]

3,5-bis(3-thienyl)-1,3,4-thiadiazole	Aerated and deaerated 1 mol/L HCl	MS[b]	0.025–0.15 mmol/L	WL, LSW and EIS at 30–60 °C	87.7–97.9% CR = 0.11–0.47 mg/cm² h	[95] and [96]
Formaldehyde: phenol (1:2 mixture)	15% HCl	N80 steel[b]	0.1–0.8% (V/V)	WL and LSW at 25–90 °C and 6–24 h IT	46.2–71.2% CR = 1.3–132.9 mm/y	[127][a,c]
Formaldehyde: o-cresol (1:2 mixture)	15% HCl	N80 steel[b]	0.1–0.8% (V/V)	WL and LSW at 25–115 °C and 6–24 h IT	50.5–93.3% CR = 0.3–145.1 mm/y	[127][a,c]
Formaldehyde: p-cresol (1:2 mixture)	15% HCl	N80 steel[b]	0.1–0.8% (V/V)	WL and LSW at 25–115 °C and 6–24 h IT	48.5–89.5% CR = 0.6–149.8 mm/y	[127][a,c]
2-(Undecyldimethyl-ammonio)butanol bromide	Deaerated 1 mol/L HCl	99.5% Fe	0.01–10 mmol/L	Tafel extrapolation at room T and WL	28–98%	[82]
2-(Dodecyldimethyl-ammonio)butanol bromide	Deaerated 1 mol/L HCl	99.5% Fe	0.01–10 mmol/L	Tafel extrapolation at room T and WL	24–99.5%	[82]
2-(Tridecyldimethyl-ammonio)butanol bromide	Deaerated 1 mol/L HCl	99.5% Fe	0.01–10 mmol/L	Tafel extrapolation at room T and WL	11–96%	[82]
2-(Tetradecyldimethyl-ammonio)butanol bromide	Deaerated 1 mol/L HCl	99.5% Fe	0.01–10 mmol/L	Tafel extrapolation at room T and WL	15–99.5%	[82]
2-(Pentadecyldimethyl-ammonio)butanol bromide	Deaerated 1 mol/L HCl	99.5% Fe	0.01–10 mmol/L	Tafel extrapolation at room T and WL	9–99%	[82]
Henna extract	Deaerated and non-deaerated 1 mol/L HCl	MS[b]	0.2–1.1 g/L	WL, LSW and EIS at 25–60 °C	9.6–92.6% CR = 0.04–1.50 mg/cm² h	[101]
2-Hydroxy-1,4-naphthoquinone (Lawsone)	Deaerated 1 mol/L HCl	MS[b]	0.2–1.1 g/L	LSW and EIS at 25 ± 1 °C	62.9–94.4%	[101]
3,4,5-Trihydroxy-benzoic acid (gallic acid)	Deaerated 1 mol/L HCl	MS[b]	0.2–1.1 g/L	LSW and EIS at 25 ± 1 °C	28.9–63.2%	[101]
-D-Glucose	Deaerated 1 mol/L HCl	MS[b]	0.2–1.1 g/L	LSW and EIS at 25 ± 1 °C	23.5–50.3%	[101]
Tannic acid	deaerated 1 mol/L HCl	MS[b]	0.2–1.1 g/L	LSW and EIS at 25 ± 1 °C	18.8–34.7%	[101]
Justicia gendarussa extract[a]	deaerated 1 mol/L HCl	MS[b]	10–200 mg/L	LSW at 25 ± 2 °C	56.7–91.3%	[102]
1-(2-Pyridylazo)-2-naphthol	open to air 1 mol/L HCl	CRS[b]	5–100 μmol/L	WL at 35–50 °C	16.0–95.2%	[83]
Tributylamine + 0.6% (w/V) formaldehyde	15% (w/V) HCl	CS UNS-G4130[b]	2% (w/V)	WL at 60 °C for 3 h	97.58%	[18] and [70]

Aniline + 0.6% (w/V) formaldehyde	15% (w/V) HCl	CS UNS-G4130[b]	2% (w/V)	WL at 60 °C for 3 h	96.66%	[18] and [70]
n-Octylamine + 0.6% (w/V) formaldehyde	15% (w/V) HCl	CS UNS-G4130[b]	2% (w/V)	WL at 60 °C for 3 h	92.33%	[18] and [70]
Diphenylamine + 0.6% (w/V) formaldehyde	15% (w/V) HCl	CS UNS-G4130[b]	2% (w/V)	WL at 60 °C for 3 h	92.04%	[18] and [70]
Dodecylamine + 0.6% (w/V) formaldehyde	15% (w/V) HCl	CS UNS-G4130[b]	2% (w/V)	WL at 60 °C for 3 h	91.37%	[18] and [70]
di-n-Butylamine + 0.6% (w/V) formaldehyde	15% (w/V) HCl	CS UNS-G4130[b]	2% (w/V)	WL at 60 °C for 3 h	91.29%	[18] and [70]
Cyclohexylamine + 0.6% (w/V) formaldehyde	15% (w/V) HCl	CS UNS-G4130[b]	2% (w/V)	WL at 60 °C for 3 h	90.32%	[18] and [70]
n-Butylamine + 0.6% (w/V) formaldehyde	15% (w/V) HCl	CS UNS-G4130[b]	2% (w/V)	WL at 60 °C for 3 h	83.84%	[18] and [70]
Triethylamine + 0.6% (w/V) formaldehyde	15% (w/V) HCl	CS UNS-G4130[b]	2% (w/V)	WL at 60 °C for 3 h	81.17%	[18] and [70]
Hexylamine + 0.6% (w/V) formaldehyde	15% (w/V) HCl	CS UNS-G4130[b]	2% (w/V)	WL at 60 °C for 3 h	75.76%	[18] and [70]
sec-Butylamine + 0.6% (w/V) formaldehyde	15% (w/V) HCl	CS UNS-G4130[b]	2% (w/V)	WL at 60 °C for 3 h	75.54%	[18] and [70]
Diethylamine + 0.6% (w/V) formaldehyde	15% (w/V) HCl	CS UNS-G4130[b]	2% (w/V)	WL at 60 °C for 3 h	74.51%	[18] and [70]
Propylamine + 0.6% (w/V) formaldehyde	15% (w/V) HCl	CS UNS-G4130[b]	2% (w/V)	WL at 60 °C for 3 h	74.2%	[18] and [70]
Isopropylamine + 0.6% (w/V) formaldehyde	15% (w/V) HCl	CS UNS-G4130[b]	2% (w/V)	WL at 60 °C for 3 h	72.42%	[18] and [70]
1,3-Dibutyl-2-thiourea + 0.6% (w/V) formaldehyde	15% (w/V) HCl	CS UNS-G4130[b]	2% (w/V)	WL at 60 °C for 3 h	95.51%	[18] and [70]
1,3-Diethyl-2-thiourea + 0.6% (w/V) formaldehyde	15% (w/V) HCl	CS UNS-G4130[b]	2% (w/V)	WL at 60 °C for 3 h	88.33%	[18] and [70]
1,3-Dimethyl-2-thiourea + 0.6% (w/V) formaldehyde	15% (w/V) HCl	CS UNS-G4130[b]	2% (w/V)	WL at 60 °C for 3 h	70.69%	[18] and [70]
Thiourea + 0.6% (w/V) formaldehyde	15% (w/V) HCl	CS UNS-G4130[b]	2% (w/V)	WL at 60 °C for 3 h	38.07%	[18] and [70]
Propargyl alcohol + 0.6% (w/V) formaldehyde	15% (w/V) HCl	CS UNS-G4130[b]	2% (w/V)	WL at 60 °C for 3 h	97.56%	[18] and [70]
2-Pentyn-1-ol + 0.6% (w/V) formaldehyde	15% (w/V) HCl	CS UNS-G4130[b]	2% (w/V)	WL at 60 °C for 3 h	97.42%	[18] and [70]
3-Butyn-1-ol + 0.6% (w/V) formaldehyde	15% (w/V) HCl	CS UNS-G4130[b]	2% (w/V)	WL at 60 °C for 3 h	97.41%	[18] and [70]
2-Butyn-1-ol + 0.6% (w/V) formaldehyde	15% (w/V) HCl	CS UNS-G4130[b]	2% (w/V)	WL at 60 °C for 3 h	95.92%	[18] and [70]
2-Butyne-1,4-diol + 0.6% (w/V) formaldehyde	15% (w/V) HCl	CS UNS-G4130[b]	2% (w/V)	WL at 60 °C for 3 h	94.41%	[18] and [70]
4-(2'-Amino-5'-methylphenylazo) antipyrine	2 mol/L HCl	MS[b]	1–10 mmol/L	WL and LSW at 30 °C	75.5–95.7%	[86]
Furfuryl alcohol	15% HCl	N80[b]	30–80 mmol/L	WL and LSW at 30–110 °C	72.0–90.2% CR = 1.2–209.6 mm/y	[36][c]
Alanine	0.1 mol/L HCl	[b]	0.1–100 mmol/L	pH = 1.16 (0.1 mol/L inhibitor), LSW at 25 °C	28.5–80.2% CR = 3.13–11.38 mm/y	[103]

Glycine	0.1 mol/L HCl	b	0.1–100 mmol/L	pH = 1.54 (0.1 mol/L inhibitor), LSW at 25 °C	60.4– 79.0% CR = 3.33– 25.4 mm/y	[103]
Leucine	0.1 mol/L HCl	b	0.1–100 mmol/L	pH = 1.58 (0.1 mol/L inhibitor), LSW at 25 °C	17.9– 91.6% CR = 1.31– 18.6 mm/y	[103]
N,N'-ortho-phenylen acetyle acetone imine	1 mol/L HCl	DIN CK45 CS[b] (perlite and martensite)	50–400 mg/L	EIS	24.9– 82.6%	[147]
4-[(3-{[1-(2-Hydroxy phenyl)methylidene] amino} propyl] ethanemidol]-1,3-benzenediol	1 mol/L HCl	DIN CK45 CS[b] (perlite and martensite)	50–400 mg/L	EIS	37.1– 55.6%	[147]
1-Ethyl-4(2,4-dinitrophenyl) thiosemicarbazide	2 mol/L HCl	CS[b]	1–16 µmol/L	WL, LSW and EIS at 30 °C	7.1– 75.0%	[33]
1,4-Diphenylthiose-micarbazide	2 mol/L HCl	CS[b]	1–16 µmol/L	WL, LSW and EIS at 30 °C	4.5– 73.6%	[33]
1-Ethyl-4-phenylthiose-micarbazide	2 mol/L HCl	CS[b]	1–16 µmol/L	WL, LSW and EIS at 30 °C	2.8– 70.6%	[33]
Quinolin-5- ylmethylene-3-{[8-(trifluoromethyl) quinolin-4-yl]thio} propanohydrazide	1 and 2 mol/L HCl	MS[b]	10–500 mg/L	WL, LSW and EIS at 30–60 °C	23.5– 93.6% CR = 1.1– 30.0 mm/y	[98]
3-Undecane-4-aryl-5-mercapto-1,2,4-triazole	15% HCl	MS[b]	500–5000 mg/L	WL at 105 ± 2 °C and LSW at 28 ± 2 °C	42.31– 83.37% CR = 5739– 8319 mm/y	[21]
3-(Heptadeca-8-ene)-4-aryl-5-mercapto-1,2,4-triazole	15% HCl	MS[b]	500–5000 mg/L	WL at 105 ± 2 °C and LSW at 28 ± 2 °C	61.23– 95.25% CR = 1365– 5591 mm/y	[21]
3-(Deca-9-ene)-4-aryl-5-mercapto-1,2,4-triazole	15% HCl	MS[b]	500–5000 mg/L	WL at 105 ± 2 °C and LSW at 28 ± 2 °C	51.42– 99.14% CR = 548– 7005 mm/y	[21]
3-(Deca-9-ene)-4-aryl-5-mercapto-1,2,4-triazole	15% HCl	N-80	500–5000 mg/L	WL at 105 ± 2 °C, IT 0.5–6.0 h and LSW at 28 ± 2 °C	57.44– 95.53 CR = 106– 475 mm/y	[21]
4-Salicylideneamino-3-hydrazino-5-mercapto-1,2,4-triazole	15% HCl	N-80[b]	250–1000 mg/L	LSW at 28 ± 2 °C	19.15– 72.34%	[65]
Propargyl alcohol	15% HCl	N-80[b]	250–1000 mg/L	LSW at 28 ± 2 °C	40.42– 53.17%	[65]
4-Salicylideneamino-3-hydrazino-5-mercapto-1,2,4-triazole	15% HCl	CRMS[b]	250–750 mg/L	LSW at 28 ± 2 °C	98.29%	[65]

Propargyl alcohol	15% HCl	CRMS[b]	250–1000 mg/L	LSW at 28 ± 2 °C	77.14–90.00%	[65]
2-Undecane-5-mercapto-1-oxa-3,4-diazole	15% HCl	CRMS[b]	500 mg/L	LSW at 28 ± 2 °C	98.94%	[87]
2-Heptadecene-5-mercapto-1-oxa-3,4-diazole	15% HCl	CRMS[b]	500 mg/L	LSW at 28 ± 2 °C	69.14%	[87]
2-Decene-5-mercapto-1-oxa-3,4-diazole	15% HCl	CRMS[b]	500 mg/L	LSW at 28 ± 2 °C	97.77%	[87]
2-Undecane-5-mercapto-1-oxa-3,4-diazole	15% HCl	N-80[b]	500 mg/L	LSW at 28 ± 2 °C	44.68%	[87]
2,4-Didimethyl aminobenzyledene aminophenylene	15% HCl	CRMS[b]	1000–5000 mg/L	WL at 105 ± 2 °C	47.6–69.3%, CR = 4424–7556 mm/y	[67]
2,4-Divanilledene aminophenylene	15% HCl	CRMS[b]	1000–5000 mg/L	WL at 105 ± 2 °C	38.4–70.3%, CR = 4280–8876 mm/y	[67]
2,4-Disalicyledene aminophenylene	15% HCl	CRMS[b]	1000–5000 mg/L	WL at 105 ± 2 °C	27.1–54.4%, CR = 6579–10517 mm/y	[67]
2,4-Dibenzyledene aminophenylene	15% HCl	CRMS[b]	1000–5000 mg/L	WL at 105 ± 2 °C	60.4–96.9%, CR = 443–5710 mm/y	[67]
2,4-Dicinnamyledene aminophenylene	15% HCl	CRMS[b]	1000–5000 mg/L	WL at 105 ± 2 °C	96.6–99.8%, CR = 36–493 mm/y	[67]
2,4-Dicinnamyledene aminophenylene	15% HCl	N-80	2000, 5000 mg/L	WL at 105 ± 2 °C	83.8–99.1%, CR = 27.4–293.4 mm/y	[67]
3,5-Diphenyl-imino-1,2,4-dithiazolidine	1 mol/L HCl	MS[b]	25–500 mg/L	LSW and WL at 25–50 °C	76.6–98.9%, CR = 0.44–3.28 mm/y	[63]
3-Phenylimino-5-chlorophenyl-imino-1,2,4-dithiazolidine	1 mol/L HCl	MS[b]	25–500 mg/L	LSW and WL at 25–50 °C	78.8–99.3%, CR = 0.29–2.34 mm/y	[63]
3-Phenyl-imino-5-tolyl-imino-1,2,4-dithiazolidine	1 mol/L HCl	MS[b]	25–500 mg/L	LSW and WL at 25–50 °C	91.7–99.7%, CR = 0.16–1.61 mm/y	[63]

3-Phenyl-imino-5-anisidylimino-1,2,4-dithiazolidine	1 mol/L HCl	MS[b]	25–500 mg/L	LSW and WL at 25–50 °C	92.6–99.8%, CR = 0.11–1.22 mm/y	[63]
1-Undecane-4-phenyl thiosemicarbazide	1 mol/L HCl	CRMS[b]	25–500 mg/L	LSW and WL at 28–65 °C	80.8–97.9%, CR = 1.3–4.3 mm/y	[88]
1-Heptadecene-4-phenyl thiosemicarbazide	1 mol/L HCl	CRMS[b]	25–500 mg/L	LSW and WL at 28–65 °C	87.1–99.4%, CR = 0.36–3.61 mm/y	[88]
1-Decene-4-phenyl thiosemicarbazide	1 mol/L HCl	CRMS[b]	25–500 mg/L	LSW and WL at 28–65 °C	89.1–99.6%, CR = 0.26–4.77 mm/y	[88]
Dicinnamylidene acetone[a]	15% HCl	N-80	1000–5000 mg/L	WL at 105 ± 2 °C	62.1–96.0%, CR = 115.0–1088.0 mm/y	[22]
Dicinnamylidene acetone + 1000 mg/L KI	15% HCl	N-80	1000 mg/L	WL at 105 ± 2 °C	92.9%	[22]
Disalicylidene acetone[a]	15% HCl	N-80	1000–5000 mg/L	WL at 105 ± 2 °C	76.0–98.7%, CR = 36.1–688.8 mm/y	[22]
Disalicylidene acetone + 1000 mg/L KI	15% HCl	N-80	1000 mg/L	WL at 105 ± 2 °C	97.7%	[22]
Divanillidene acetone[a]	15% HCl	N-80	1000–5000 mg/L	WL at 105 ± 2 °C	14.4–29.8%, CR = 2018.1–2460.8 mm/y	[22]
Divanillidene acetone + 1000 mg/L KI	15% HCl	N-80	1000 mg/L	WL at 105 ± 2 °C	21.9%	[22]
Trans-cinnamaldehyde	15% HCl	API J55[b]	0.015 mol/L	WL at 65 °C	91.9%	[108]
Trans-cinnamaldehyde + (0.0015 mol/L) n-dodecylpyridinium bromide	1–20% HCl	API J55[b]	0.03 mol/L trans-cinnamaldehyde	WL at 29.4–93.9 °C	CR = 0.026–1.541 kg/m²/day	[108]
2-Benzoyl-3-hydroxy-1-propene	15% HCl	J55	2 g/L	WL at 65 °C	91.6%	[27]
2-Benzoyl-3-hydroxy-1-propene + adduct of trimethyl-1-heptanol with 7 mol of ethylene oxide (THEO)	15–28% HCl	J55	2–4 g/L + 0.5–1.0 g/L THEO	WL at 65 °C	99.2–99.3%	[27]
2-Benzoyl-3-hydroxy-1-propene + N-dodecylpyridinium bromide (DDPB)	15–28% HCl	J55	2–4 g/L + 0.5–1.0 g/L DDPB	WL at 65 °C	98.5–99.1%	[27]

2-Benzoyl-3-methoxy-1-propene	15% HCl	J55	2 g/L	WL at 65 °C	94.7%	[27]
2-Benzoyl-3-methoxy-1-propene + THEO	15–28% HCl	J55	2–4 g/L + 0.5–1.0 g/L THEO	WL at 65 °C	99.0–99.2%	[27]
2-Benzoyl-3-methoxy-1-propene + DDPB	15–28% HCl	J55	2–4 g/L + 0.5–1.0 g/L DDPB	WL at 65 °C	98.8–99.0%	[27]
5-Benzoyl-1,3-dioxane	15% HCl	J55	2 g/L	WL at 65 °C	56.6%	[27]
5-Benzoyl-1,3-dioxane + THEO	15–28% HCl	J55	2–4 g/L + 0.5–1.0 g/L THEO	WL at 65 °C	84.0–98.9%	[27]
5-Benzoyl-1,3-dioxane + DDPB	15–28% HCl	J55	2–4 g/L + 0.5–1.0 g/L DDPB	WL at 65 °C	94.5–98.6%	[27]
2-Benzoyl-1,3-dimethoxypropane	15% HCl	J55	2 g/L	WL at 65 °C	60.4%	[27]
2-Benzoyl-1,3-dimethoxypropane + THEO	15–28% HCl	J55	2–4 g/L + 0.5–1.0 g/L THEO	WL at 65 °C	90.7–99.1%	[27]
2-Benzoyl-1,3-dimethoxypropane + DDPB	15–28% HCl	J55	2–4 g/L + 0.5–1.0 g/L DDPB	WL at 65 °C	97.5–99.1%	[27]
3-Hydroxy-1-phenyl-1-propanone	15% HCl	J55	2 g/L	WL at 65 °C	0%	[27]
3-Hydroxy-1-phenyl-1-propanone + THEO	15% HCl	J55	2 g/L + 0.5 g/L THEO	WL at 65 °C	98.8%	[27]
3-Hydroxy-1-phenyl-1-propanone + DDPB	15% HCl	J55	2 g/L + 0.5 g/L DDPB	WL at 65 °C	98.5%	[27]
1-(2-Ethylamino)-2-methylimidazoline	Deaerated 0.5 mol/L HCl	CS[b]	10–100 mg/L	LSW and EIS at room T	39–70%	[84]
N-[3-(2-Amino-ethylaminoethyl)]-acetamide	Deaerated 0.5 mol/L HCl	CS[b]	10–100 mg/L	LSW and EIS at room T	44–77%	[84]
1-(2-Ethylamino)-2-methylimidazolidine	Deaerated 0.5 mol/L HCl	CS[b]	10–100 mg/L	LSW and EIS at room T	0–20%	[84]
1-(2-Aminoethyl)-2-oleylimidazoline	15% HCl	N80[b]	10–150 mg/L	WL, LSW, EIS at 25–50 °C	68.99–96.23% CR = 0.26–2.93 mm/y	[17]
1-(2-Oleylamidoethyl)-2-oleylimidazoline	15% HCl	N80[b]	10–150 mg/L	WL, LSW, EIS at 25–50 °C	64.25–91.16% CR = 0.93–3.41 mm/y	[17]
1-Cinnamylidine-3-thiocarbohydrazide	15% HCl	CS[b]	500–2000 mg/L	WL, LSW, EIS, hydrogen permeation current measurements at 30–110 °C	87.4–98.5% CR = 0.71–255.96 mm/y	[99]
1,1′-Dicinnamylidine-3-thiocarbohydrazide	15% HCl	CS[b]	500–2000 mg/L	WL, LSW, EIS, hydrogen permeation current measurements at 30–110 °C	90.8–99.2% CR = 0.39–140.31 mm/y	[99]

As seen in Table 2, the most common techniques for the evaluation of CI performance are weight-loss (WL), linear sweep voltammetry – LSW (polarization resistance (R_p) or even more frequently Tafel plot measurements), and electrochemical impedance spectroscopy (EIS). The main focus below is on evaluation and not on reporting the manner of inhibitor bonding or examining adsorption isotherms.

Corrosion Inhibitors for Hydrochloric Acid Solutions

As mentioned above, among acid solutions, HCl (at 5–28% [24]) is the most widely used for the acidizing procedure and that is why the main focus is on this acid (see below and Table 2). However, not all CIs tested in HCl solutions (given below) have already been used for oilfield applications, but the emphasis herein is on summarizing what kind of compounds have already been tested in HCl and could potentially be used. That is why HCl concentrations lower than the minimum commonly employed (5%) are also included. It must be pointed out that the following review cannot cover all aspects of CI use in HCl solutions.

Ita and Offiong [68] studied benzoin and benzil compounds as CIs for MS in HCl solution at 30 and 40 °C. It was reported that the of these compounds has the following order: benzoin > benzoin-(4-phenylthiosemicarbazone) > benzil > benzil-(4-phenylthiosemicarbazone). They related this trend to the compound solubility and to the strength of the inhibitor–metal bond.

Flores et al. [69] showed that the effectiveness of sodium N-alkyl-phthalamates (alkyl = n-C_6H_{13}, n-$C_{10}H_{21}$, n-$C_{14}H_{29}$) as CIs for SAE 1018 CS in 0.5 mol/L HCl is dependent on the alkyl chain length and concentration, i.e. by increasing them increased. All there inhibitors acted as mixed-type inhibitors. The authors suggested a physisorption type of adsorption and that phthalamates form complexes and chelates with iron, which prevents iron oxidation and consequently corrosion.

Baddini et al. [18] and Cardoso et al. [70] studied 23 compounds as CIs for UNS-G4130 CS in 15% (w/V) HCl at 60 °C. Along with active inhibitor compounds, they employed formaldehyde to minimize hydrogen penetration into steel. Baddini et al. [18] found that tributylamine, some alcohols (Table 2), aniline, n-octylamine, diphenylamine, dodecylamine, di-*n*-butylamine, cyclohexylamine, 1, 3-dibutyl-2-thiourea are the most effective CIs among the compounds studied.

Vishwanatham and Haldar [36] reported that furfuryl alcohol is an effective mixed-type CI, with the predominant effect on the cathodic reaction, for N80 steel in 15% HCl. Its   increased with increasing CI concentration, but decreased with increasing T (temperature) from 30 to 110 °C.

Jayaperumal [29] reported that octyl alcohol and propargyl alcohol are excellent inhibitors for MS in 15% HCl at 30 and 105 °C. Especially in solution containing propargyl alcohol, as the CR was reported to be 0.4 mm/y (at 30 °C) and 3 mm/y (at 105 °C). However, according to the EIS spectra, the charge transfer resistance is relatively low (290 Ω cm^2 for the highest inhibitor concentration), indicating quite a high CR, which could raise some doubts regarding the inhibitor performance. Moreover, no explanation of how the electrochemical measurements were carried out at 105 °C is given in the text.

Babic-Samardzija et al. [71] investigated 2-butyn-1-ol, 3-butyn-1-ol, 3-pentyn-1-ol, and 4-pentyn-1-ol as CIs for iron in 1 mol/L HCl at ambient temperatures. They reported that all compounds act as mixed-type inhibitors and that their   depends on the chain length and the position of the triple bond.

Popova   et   al.   [72]   investigated   benzimidazole, 2-aminobenzimidazole, 2-mercaptobenzimidazole, 1-benzylbenzi-midazol, and 1, 2-dibenzylbenzimidazole as CIs for MS in deaerated 1 mol/L HCl solution. They reported that all five diazoles have pronounced corrosion inhibition properties, whereas the latter three were particularly effective. Subsequently, Popova et al. [73] reported that 8 benzimidazole derivatives exhibit a corrosion inhibition effect for MS in 1 mol/L HCl at 20 °C. Their   increases with increased CI concentration and they mainly act as mixed-type inhibitors. The   of these compounds has the following order: 5(6)-nitrobenzimidazole < benzimidazole<2-methylbenzimidazole<5(6)-carboxybenzimidazole < 2-hydroxymethylbenzimidazole < 2-aminobenzimidazole ≈ 2-benzimidazolylacetonitrile < 2-mercaptobenzimidazole. However, the reported R_p values from the LSW measurements are very low (in all cases less than 1 Ω cm^2), which raises some doubt about the effectiveness of the compounds studied. Most likely the measured R_p

values were in $k\Omega$ cm^2 and there is simply a typographical error in the article. Furthermore, Popova et al. [74] investigated 5 different azole compounds as CIs for MS in 1 mol/L HCl at 20 °C. The   trend of these compounds was reported to have the following order: indole ≈ 1H-benzotriazole ≈ benzothiazole > benzimidazole. On the other hand, the addition of benzothiadiazole to the HCl solution even promoted CR compared with the non-inhibited solution. The authors also reported that the   of these compounds, except benzothiadiazole, increases with increased CI concentration and that they act as mixed-type inhibitors by predominantly reducing the rate of the anodic reaction.

Aljourani et al. [75] showed that the   trend of different CIs has the following order: 2-mercaptobenzimidazole > 2-methylbenzimidazole > benzimidazole for MS in 1 mol/L HCl at 25 °C. The   of all 3 CIs decreased with increasing T from 25 to 55 °C. Moreover, it has to be pointed out that their   calculation from the experimental results obtained by means of LSW and EIS techniques gave quite large difference (up to approximately 20% for the highest CIs concentration).

Khaled [76], Khaled and Hackerman [77], and Khaled et al. [78] studied the corrosion inhibition of pure Fe in 1 mol/L HCl at 25 °C. In the first study, wherein Khaled [76] examined different benzimidazoles, he showed that the   trend of these compounds has the following order: 2-aminobenzimidazole > 2-(2-pyridyl)benzimidazole > 2-aminomethylbenzimidazole > 2-hydroxybenzimidazole > benzimidazole. In a study on o-substituted anilines [77] they showed that the   trend has the following order: 2-chloroaniline > 2-fluoroaniline > 2-methoxyaniline ≈ 2-ethoxyaniline > 2-ethylaniline > 2-methylaniline. For piperidine and its 6 derivatives [78], the authors reported that the inhibition performance has the following order: cis-2, 6-dimethylpiperidine < 3, 5-dimethylpiperidine < 2-methylpiperidine < 3-methylpiperidine < piperidine < 4-benzylpiperidine < 4-methylpiperidine. Finally, it was shown that in each case all compounds acted as mixed-type inhibitors.

Cruz et al. [79] investigated 2-aminomethylbenzimidazole and bis(benzimidazol-2-ylethyl)sulphide as CIs for CS in deaerated 0.5 mol/L HCl. They reported that the former acts as a cathodic-type inhibitor and the latter as a mixed-type inhibitor.

Ait Chikh et al. [80] studied the adsorption and inhibition properties of 1,12-bis(1,2,4-triazolyl)dodecane for CS in 1 mol/L HCl. This compound acted as a good cathodic-type inhibitor. They suggested that adsorption of this compound occurs via synergistic effect between chloride ions and the positive quaternary ammonium ion moiety present in the inhibitor molecule. However, the inhibition effect diminished in aerated compared with deaerated solution and with prolonged immersion time of the CS in HCl solution.

Abiola [81] reported that 3-(4-amino-2-methyl-5-pyrimidyl methyl)-4-methyl thiazolium chloride effectively prevents hydrogen evolution and corrosion of MS in 0.5 mol/L and 5 mol/L HCl at 30 °C.

Elachouri et al. [82] employed known surfactants, i.e. some 2-(alkyl(C_nH_{n+1})dimethylamonio)butanol bromides ($n = 11-15$) as CIs for Fe (purity 99.5%) and showed that they are effective cathodic-type CIs. Their increased with an increased number of C atoms in the side chain and an increased Cl concentration.

Tang et al. [83] reported that 1-(2-pyridylazo)-2-naphthol is an effective mixed-type CI, with a predominant inhibition effect on the anodic reaction, for CS in 1 mol/L HCl at 25–50 °C. Moreover, its decreases with increasing T.

Cruz et al. [84] showed that the trend of the η has the following order: 1-(2-ethylamino)-2-methylimidazoline $\gg$ N-[3-(2-amino-ethylaminoethyl)]-acetamide > 1-(2-ethylamino)-2-methylimidazoli-dine for CS at room T in deaerated 0.5 mol/L HCl.

Ita and Offiong [85] reported that -pyridoin is more effective than 2, 2'-pyridil as a CI for MS (only a composition of 98% iron was specified) in 0.5 mol/L HCl at 30 and 40 °C. The author also showed that these compounds prevent hydrogen evolution in 8 mol/L HCl. Moreover, the of these CIs increases with increased concentration and with increasing T.

The Application of Newly Synthesized Compounds in HCl Solutions

Researchers attempt to increase the environmental acceptability of potential CI compounds by synthesizing new chemicals. Some examples are given below.

AbdElRehimetal. [86] synthesized 4-(2'-amino-5'-methylphenylazo) antipyrine and tested it as a CI for MS in 2 mol/L HCl at 20–60 °C. They reported that this compound is an effective mixed-type inhibitor and its increases with increased concentration, but decreases with increasing T.

Quraishi and Jamal have published numerous studies on this field. They synthesized 3 fatty acid triazoles[21], i.e. 3-undecane-4-aryl-5-mercapto-1,2,4-triazole, 3(heptadeca-8-ene)-4-aryl-5-mercapto-1,2,4-triazole, and 3(deca-9-ene)-4-aryl-5-mercapto-1,2-4-triazole, and tested them as CIs for CRMS (cold rolled mild steel) and N-80 steel in 15% HCl at 28 ± 2°C and boiling T (105 ± 2°C). The authors claim that these compounds are environmentally benign and have low toxicity. They showed that all compounds act as effective mixed-type CIs and their increases with increased inhibitor concentration. Subsequently, Quraishi and Jamal [65] synthesized another compound, i.e. 4-salicylideneamino-3-hydrazino-5-mercapto-1, 2, 4-triazole, which they claim is eco-friendly and cost efficient. The of this compound was compared with propargyl alcohol in 15% HCl at 28 ± 2°C and 105 ± 2 °C for CRMS and N-80 steel. Propargyl alcohol was used as a standard CI for the acidization process. The authors reported that the triazole compound acts as a mixed-type inhibitor, whereas propargyl alcohol acts as an anodic-type inhibitor for CRMS and as a cathodic-type inhibitor for N-80 steel. Moreover, propargyl alcohol was more efficient for both steel materials at concentrations above 750 mg/L. On the other hand, the authors stated that the triazole compound does not produce toxic vapours like propargyl alcohol during the acidization process. Furthermore, Quraishi and Jamal [87] synthesized three oxadiazoles, i.e. 2-undecane-5-mercapto-1-oxa-3, 4-diazole, 2-heptadecene-5-mercapto-1-oxa-3, 4-diazole and 2-decene-5-mercapto-1-oxa-3,4-diazole and tested them as CIs for MS in 15% HCl at 28 ± 2°C and 105 ± 2°C. The authors reported that all these compounds act as effective mixed-type CIs and that 2-undecane-5-mercapto-1-oxa-3, 4-diazole is the most effective among them. Moreover, it was shown that 2-undecane-5-mercapto-1-oxa-3, 4-diazole under the same experimental conditions is an effective CI for N-80 steel. Next, Quraishi and Jamal [67] synthesized different dianils as condensation products of aromatic aldehydes and p-phenylenediamine, i.e. 2,4-didimethyl aminobenzyledene aminophenylene (DDAP), 2,4-divanilledene aminophenylene (DVAP),

2,4-disalicyledene aminophenylene (DSAP), 2,4-dibenzyledene aminophenylene (DBAP), and 2,4-dicinnamyledene aminophenylene (DCAP), and tested them as CIs for CRMS 15% HCl at 105 ± 2°C. They reported that the   at 3000 mg/L CI concentration has the following order: DCAP > DBAP > DVAP > DDAP > DSAP. Moreover, all these CIs acted as mixed-type inhibitors at 28 ± 2°C. DCAP was also tested for N-80 steel in the same solution and the authors reported that it behaves predominantly as an anodic-type CI at 28 ± 2°C. At 105 ± 2 C, DCAP exhibited a high , whereas its decreased with increased immersion time from 0.5 h to 6 h. Quraishi and Sardar [63] also synthesized different dithiazolidines, i.e. 3,5-diphenyl-imino-1,2,4-dithiazolidine (DPID), 3-phenyl-imino-5-chlorophenyl-imino-1,2,4-dithiazolidine (PCID), 3-phenyl-imino-5-tolyl-imino-1,2,4-dithiazolidine (PTID), and 3-phenyl-imino-5-anisidylimino-1,2,4-dithiazolidine (PAID), and tested them as CIs for MS in 1 mol/L HCl. The of all CIs increases with increased concentration and has the following order: PAID > PTID > PCID > DPID at 25°C. Furthermore, Quarishi et al. [88] synthesized different thiosemicarbazides of fatty acid, i.e. 1-undecane-4-phenyl thiosemicarbazide, 1-heptadecene-4-phenyl thiosemicarbazide, and 1-decene-4-phenyl thiosemicarbazide, and tested them as CIs for CRMS in 1 mol/L HCl. They reported that the of the CIs increases with increased concentration at 35°C and does not chance significantly by prolonging the immersion time and by raising the T from 35°C to 65°C, when a 500 mg/L concentration of each CI was employed. Quraishi et al. [22] also synthesized dicinnamylidene acetone, disalicylidene acetone, and divanillidene acetone, and tested them as CIs for N-80 steel in 15% HCl. The first two compounds were more effective than the latter at 105 ± 2°C. For all three compounds, the presence of KI as an intensifier increased their . The authors also reported that all three compounds act as mixed-type inhibitors in 15% HCl at 40 ± 2°C.

Yildirim and Cetin [89] synthesised different acetamide, isoxazolidine, and isoxazoline derivatives with a long alkyl side chain and tested them as CIs for cold rolled low CS DIN EN 10130-99 in 2 mol/L HCl containing 10% acetone at room T. They showed that newly synthesized compounds act as very effective CIs and almost all exhibit the highest η at 50 mg/L concentration. The disadvantage of the presented results for practical use in oil-field applications is that the inhibitors were introduced into the acid medium in 10% acetone and that corrosion products after the test were removed by emery paper (even though this was done gently).

Wang et al. [90] studied corrosion inhibition of MS at 25°C in 1 mol/L HCl by four synthesized mercapto-triazole compounds, which acted as efficient mixed-type inhibitors.

Ali et al. [91] synthesized different isoxazolidines. In 1 mol/L HCl solution at 60°C they examined the corrosion inhibition properties of these compounds for two MS types and found good performance. They claim that the presence of adjacent heteroatoms (N–O) with three lone pairs of electrons in isoxazolidine moiety invariably plays a dominant role in corrosion inhibition. However, even though the   values are high, the reported corrosion current densities in inhibited solutions are high and consequently CR is suspected to be high as well.

Mernari et al. [92] synthesized different 3, 5-bis(n-pyridyl)-4-amino-1,2,4-triazoles ($n = 1, 2, 3$) and tested their   for MS in 1 mol/L HCl at 30 C. They showed that these compounds act as effective anodic-type inhibitors whose   increases with increased inhibitor concentration. Subsequently, some members of this research group showed that 2,5-diphenyl-1,3,4-oxadiazole, 3,5-diphenyl-1,2,4-triazoles 2,5-di(n-pyridyl)-1,3,4-oxadiazoles [93], and 1,4-bis(2-pyridyl)-5H-pyridazino [4,5-b]indole [94] could act as CIs for MS in 1 mol/L HCl at 30 C. Furthermore, Bentiss et al. [95] and Lebrini et al. [96] used 2 compounds considered to be non-cytotoxic substances as CIs for MS in 1 mol/L HCl, i.e. 3,5-bis(2-thienyl)-1,3,4-thiadiazole and 3,5-bis(3-thienyl)-1,3,4-thiadiazole. They showed that these compounds are mixed-type inhibitors with a predominant inhibition effect on the cathodic reaction. Moreover, their   increases with increased concentration and T.

Qui et al. [97] synthesized 3 gemini surfactants (these contain two hydrophilic groups and two hydrophobic groups in the molecule), i.e. 1, 2-ethane bis(dimethyl alkyl (C_nH_{2n+1}) ammonium bromide) ($n = 10$, 12, 16) and tested them as CIs for A_3 CS in 1 mol/L HCl at 25 C. The authors showed that these compounds are effective in preventing corrosion and their   increases with increased surfactant concentration, reaching the maximum value near the critical micelle concentration.

Badr [33] synthesized 3 thiosemicarbazide compounds and tested them as CIs for CS in 2 mol/L HCl at 30°C. They reported that the   of these compounds has the following order: 1-ethyl-4(2, 4-dinitrophenyl) thiosemicarbazide > 1, 4-diphenylthiosemicarbazide > 1-ethyl-4-phenylthiosemicarbazide, that they all act as mixed-type inhibitors, and that their   increases with increased concentration.

Saliyan and Adhikari [98] synthesized quinolin-5-ylmethylene-3-{[8-(trifluoromethyl)quinolin-4-yl]thio}propanohydrazide and tested it as a CI for MS in 1 and 2 mol/L HCl at 30–60°C. They reported that this compound acts as an anodic-type CI, and that its   increases with increased concentration and slightly decreases with increasing T.

Yadav et al. [17] synthesized 1-(2-aminoethyl)-2-oleylimidazoline and 1-(2-oleylamidoethyl)-2-oleylimidazoline and tested them as CIs for N80 steel in 15% HCl. They showed that both compounds act as mixed-type inhibitors at 25°C and that the CR of N80 steel increases with increasing T from 25°C to 50°C, when both compounds are present at a concentration of 150 mg/L.

Sathiya Priya et al. [99] synthesized 1-cinnamylidine-3-thiocarbohydrazide and 1, 1'-dicinnamylidine-3-thiocarbohydrazide and tested them as CIs for CS in 15% HCl. They showed that these compounds act as mixed-type inhibitors at 30°C. When T increased from 30°C to 110°C, the CR increased, but the   remained between 98.2% and 99.1%. The authors also showed that these compounds effectively decrease the hydrogen permeation current compared with non-inhibited solution.

Natural Products as Corrosion Inhibitors in HCl Solutions

The hazards of most synthetic organic inhibitors is commonly known and the restrictive environmental regulations of many countries forced researchers to focus on developing cheap, non-toxic, and environmentally acceptable products. Due to this reason, some researchers suggest using plant extracts as CIs, however the resulting corrosion inhibition effectiveness is usually found to be very low. Raja and Sethuraman [100] examined some of these compounds in their review article. However, they concluded that a phytochemical investigation of the extract is rarely carried out and seldom is it known which active ingredient present in the plant extract is responsible for corrosion inhibition. Therefore, it is also likely that a mixture of constituents are acting as a CI [100]. Some examples are given below.

Ostovari et al. [101] investigated the corrosion inhibition performance of henna extract (Lawsonia inermis) and its main constituents (Lawsone, gallic acid, -D-glucose, and tannic acid) for MS in 1 mol/L HCl. Henna extract is considered a low cost, eco-friendly, and naturally occurring substance. The authors showed that this extract is effective in preventing corrosion (also pitting), however, as the T increased from 25 to 60°C the of henna extract decreased. Moreover, they tested the extract's constituents and concluded that all compounds act as mixed-type inhibitors and some of them also as oxygen scavengers. They also reported that the of these compounds has the following order: Lawsone > henna extract > gallic acid > -D-glucose > tannic acid.

Satapathy et al. [102] tested Justicia gendarussa plant extract as a CI for MS in 1 mol/L HCl at 25–70°C. They claim that the major components in this extract are -sitosterol, friedelin, lupenol, phenolic dimmers,o-substituted aromatic amines (2-amino benzyl alcohol, 2-(2'-amino-benzylamino) benzyl alcohol), and flavonoids (6,8-di-C- -L-arabinopyranosyl-4',5,7-trihydroxyflavone,   6,8-di-C- -L-arabinosylapigenine,   6-C- -L-arabinopyranosyl-4',5,7-trihydroxy-8-C- -D-xylopiranosyl-flavone,   and   6-C- -L-arabinosyl-8-C- -D-xylosylapigenine). The authors reported that this extract acts as a mixed-type inhibitor. With increasing concentration, the increases; on the other hand, the decreases with increasing T. They also reported that at 80°C it has little or no corrosion inhibition effect due to the decomposition of the extract's compounds.

Ashassi-Sorkhabi et al. [103] studied the corrosion inhibition effect of 3 amino acids, i.e. alanine, glycine, and leucine for steel (the type of steel was not mentioned; see the composition in Table 1) in HCl solutions. The authors used these compounds as they are non-toxic, relatively cheap, and easy to produce with purities greater than 99%. These 3 amino acids acted as efficient CIs, but only if the inhibitor concentration was 1 mmol/L or higher for alanine and glycine and 10 mmol/L or higher for glycine. Otherwise, the authors observed a corrosion promotion effect, most likely due to complexation with Fe.

DESIGNING THE CORROSION INHIBITOR FORMULATION

One of the most important aspects in formulating the acid system is the need to ensure adequate corrosion inhibition while providing the desired reactivity with the formation material for stimulation purposes [19]. Organic compounds alone are usually not effective enough for corrosion control and a proper mixture containing additional intensifiers, surfactants, solvents, and co-solvents is needed [39]. This mixture is then called the corrosion inhibitor formulation (CIF). Some authors refer to CIF as an inhibitor cocktail [9] or corrosion inhibitor package – CIP [104]. Hereinafter, the CIF term will be used. On the other hand, most of the literature concerns single compounds as CIs for steel materials (see Table 2), however as single compounds, they are usually not effective enough in industrial applications.

In general, the CIF has several criteria for application. As an acidizing inhibitor, the CIF must be stable (dispersive – not separated) in the acid for at least 24–72 h, which is the duration of time the acid/CIF is stored onsite. Additionally, it needs to be liquid over a wide temperature range for field use in both cold and hot climates and there should not be any separation or solidification issues. Other criteria that are often required include the pour point (−20°C), shelf life (1 year), viscosity, and other additive and H_2S compatibility requirements. The most important factors that have to be taken into account in CIF design are performance criteria, these include performance at various temperatures and pressures, exposure time, steel metallurgy, acid type and concentration, and the surfactant used. The performance cost of a particular CIF can also be a very important decision factor. Smith et al. [24] reported that the CI cost needed in the acidizing process can be compared with the cost of the pipe, therefore care should be taken when deciding which and how much of the CI is necessary. Commonly, the corrosion inhibition effectiveness is judged from the material mass loss after a certain time at a given temperature. The authors would like to note that sometimes a data comparison can be misleading due to the different testing procedures employed [24].

Finding an effective CIF is a difficult task. Usually this is done by determining the corrosion inhibition effectiveness of numerous single compounds. If they perform as effective CIs, they are often used to develop a complex mixture together with other chemicals. The goal is to improve the CIF's as compared with a single CI. Therefore, in most cases this is performed by trial and error experiments on the basis of previous knowledge. Moreover, CIF development must include a balance between environmental impact, cost, safety, and technical requirements [20] and [105].

In order to follow industrial recommendations strictly, no more than 2% w/V of active components is allowed for matrix acidification operations [18] and [30]. However, sometimes a CI concentration of up to 4% is used [27]. A commonly employed acceptable CR limit of the tested materials is 0.243 kg/m^2/test period and the pitting index of the tested material should not be higher than 3 (Table 3) [9], [19], [28] and [40]. Brondel et al.[23] reported that a CR of 2.4–9.8 kg/m^2/year without pitting is acceptable. A comprehensive study in 1978 of how CIF should be designed was given by Smith et al. [24].

Table 3: Definition of the pitting index [19], [20] and [24]

Description	Pitting index
No pits. The surface is the same as for the original untreated coupon	0
Intergranular corrosion on the cut edge of the coupon, giving a sintered effect; no pits on major surfaces	1
Small, shallow pits on cut edges: no pits on major surfaces	2
Scattered, very shallow pinpoint pits, less than 25 pits on either surface – i.e. on front or back	3
More than 25 pits of Rank 3 on either surface	4
Ten or fewer pits, 1/32- to 1/16-in. diameter, 1/64- to 1/32-in. deep	5
11–25 Pits of Rank 5	6
More than 25 pits of Rank 5	7
Pits larger than 1/16 in., but less than 1/8 in. in diameter, greater than 1/32-in. deep, 100 or fewer in number	8
Any pitting more severe than Rank 8	9

In general, a typical composition of a formulated inhibitor package contains all or some of the following components: active inhibitor substance(s), surfactant, solvent, and intensifier. Depending on the application, other additives are also sometimes added. Active inhibitor substances are mainly responsible for the inhibition of metal corrosion.

Active Corrosion Inhibitor Substances

As discussed below in Section 5.6, the most common compounds in effective CIFs are acetylenic alcohols, , -unsaturated aldehydes, -alkenylphenones, quaternary amines, and derivatives of pyridinium and quinolinium salts.

However, numerous compounds used as CIs for lower-grade steel materials are presented above and inTable 2, which could potentially be used as active corrosion inhibitor substances and formulated with surfactants, solvents, and intensifiers in order to develop an effective CIF (see below).

Surfactants

A surfactant is a surface active agent. In this work a surfactant term will be used for compounds which improve the dispersability of the CI in the acid (as emulsifiers providing dispersed emulsion – not separated) while wetting the surface of the metallic material [14], [20] and [24]. However, surfactants can offer corrosion protection themselves. Some examples when the same compound was used as a surfactant or active corrosion inhibitor ingredient are given below. Typical surfactants in the oilfield services industry are alkylphenol ethoxylates, e.g. nonylphenol ethoxylate (NPE) [14], [15], [30], [106] and [107]. However, NPEs have been banned from use in the North Sea because of their toxicity. On the other hand, ethoxylated linear alcohols are more acceptable [20]. The quaternary ammonium salts and amines (when protonated) are the most used compounds of the cationic surfactants class, where the cation is the surface active specie. As the amines only function as a surfactant in the protonated state, they cannot be used at high pH. On the other hand, quaternary ammonium compounds, frequently abbreviated as "quats", are not pH sensitive. Long-chain quaternary ammonium bromides were also reported to work as efficient CIs for steel materials[106]. A frequently employed surfactant was N-dodecylpyridinium bromide (DDPB) [9], [60], [61],[108] and [109]. Anionic sulphates, anionic sulphonates, alkoxylated alkylphenol resins, and polyoxyethylene sorbitan oleates are also useful surfactants. Ali reported that a particularly useful surfactant is a blend of polyethylene glycol esters of fatty acids and ethoxylated alkylphenols [15]. Several examples of the surfactants used are given below in Section 5.6.

Solvents

Solvents are mainly used for two purposes: to reduce viscosity for ease of handling and to ensure formulation stability in various environments. Moreover, solvents have a similar purpose as surfactants, but with a different mechanism, i.e. to improve CIF solubility and dispersability in the acid, and wettability on the acid–steel interface. Flammability is an important factor when selecting a solvent for CIF in some regions, but not all. On the other hand, we note that cost is a larger factor than flammability in solvent selection.

Hill and Romijn [20] reported that the usually employed solvents are toluene, xylem, and other aromatic solvent mixtures, however they are classified as products that cause tainting. Therefore they would need to be replaced in the future. We note that methanol and isopropanol are commonly used solvents. Methanol is a very cost effective solvent, however it is a cumulative poison [20]. OSPARCOM (Oslo Paris Commission, see below) [110] accepts it for application as it poses little or no risk to the environment. Isopropanol is an excellent solvent, but it has a low flash point and it needs to be labelled as flammable [20]. Moreover, low-molecular weight alcohols, glycols, dimethylsulphoxide, dimethylacetamide, 1-methyl-2-pyrrolidone, tetramethylene sulphone [15] and [111], formic acid, and formic acid derivatives such as dimethylformamide were also applied (the latter is classified as a mammalian mutagen). Sometimes co-solvents are also employed. An example of a co-solvent is HAN – heavy aromatic naphtha (a mixture of mainly C_9 and C_{10} aromatic hydrocarbons – predominantly trimethyl benzenes, diethyl benzenes, and dimethyl ethyl benzenes) [112], which has oil-wetting characteristics [28]. Occasionally, alcohols or glycol ethers (ethylene–glycol-monobutyl ether – EGMBE, sometimes called a mutual solvent [30] and [39]), are added to the acid to improve acid penetration and clean-up [19]. However, they can reduce the CIs effectiveness, but lower parasitic consumption. A mutual solvent is usually described as a chemical additive soluble in oil, water, and acid treatment fluids.

Intensifiers

An intensifier (sometimes called an inhibitor aid [19]) is usually added to the CIF, because organic CIs frequently cannot provide adequate protection to steels at high temperatures and long exposure times [113]. Common intensifiers include formic acid (used from 0.5 to 10 wt.% [62]; Brondel et al. [23] reported 9% for deep sour wells), methyl formate, KI [22] and [114] (which can be used from 0.1 to 2.0 wt.% [62]), CuI [15],[16], [40], [64] and [111], CuCl [15] (when the of the CI is low, Cu plating on the tubular occurs [10]), and metals ions (e.g. from Sb_2O_3, $SbCl_3$, Sb_2O_5, $K_4Sb_2O_7$, $K_2H_2Sb_2O_7$, Sb_2S_3, $SbCl_5$, $K_2Sb_2(C_4O_6H_4)_2$, $Sb[(CH_2OH)_2]_3$ [13], [16], [28], [64] and [111], $BiCl_3$, BiI_3, $BiOCl$, Bi_2O_3, $BiOI_3$, BiF_3, bismuth tartrate, bismuth adduct of ethylene glycol and bismuth trioxide, bismuth subsalicylate [14], [62], [64],[111] and [115], $SnCl_2$ [40], As^{3+}, Cr^{6+}, Cu^{2+}, Ni^{2+}, Sn^{2+}, Hg^{2+} [116], calcium salts [64], and $MgCl_2$ [111]).

It was reported that a 50:50 mixture of CuI and CuCl was much more effective than the use of individual components as an intensifiers [15]. Moreover, the reaction of insoluble Sb_2O_3 and Bi_2O_3 with HCl leads to the formation of soluble $SbCl_3$ and $BiCl_3$, respectively [64], which can then act as intensifiers. It was also reported that Bi_2O_3 is an especially effective intensifier in combination with KI [40] and [62]. Sometimes calcium chloride or bromide and zinc bromide may be used at concentrations starting at 0.1% up to saturation [27]. Williams et al. [111] also reported the possible use of Ca, Al, Mg, Zn, and Zr ions. Furthermore, formamide or formic ester have also been employed [62]. Frenier [25] reported that propionic, propiolic, acetic, and chloroacetic acids, HI, and NaI can be used as intensifiers. Hill and Jones [19] employed an antimony salt intensifier for SM25Cr steel.

Hill and Romijn [20] emphasized the need for a formic acid intensifier at T above 93 °C. However, lately it has been recommended to avoid formic acid in the CIF design due to the pipeline corrosion problems associated with its use [15]. Keeney and Johnson [40] claim that CuI significantly increases the of ethyl octynol. Williams et al. [14] and [64] claim that the function of the metal compound is to produce metal ions which form complexes (a coordination or association of the metal) with, for example, the quaternary ammonium compound, and form a protective deposit on the metal tubulars and equipment. On the other hand, they claim that Sb compounds are toxic, whereas Bi compounds have lower toxicity. Moreover, Williams et al. [64] claim that the concentration of the metal ions as intensifiers is preferred to be in the range of 1–1.5% of the total acid solution (with CIF inside the acid), due to economic reasons. It is also interesting that Hill and Jones [19] used less CI (1.2%) compared with the intensifier (5%) to protect steel at 163°C. The authors also claim that formic acid, antimony salts, and KI are acceptable intensifiers for the CIF design to be used in the North Sea.

Other Additives

Commonly, also other additives, along with the above-mentioned substances, are added to the acidizing fluids. They do not have the purpose of inhibiting corrosion, but they can influence the corrosion inhibition performance of the known CIFs and increase the CR significantly. These include iron control agents, water wetting agents, anti-sludge agents, non-emulsifiers, stabilizers, and viscoelastic surfactants.

Ferric iron forms a gelatinous mass precipitate, which prevents or slows down the flow through the channels. This precipitated iron plug therefore decreases production. Iron control agents are used to prevent this issue. The source of iron comes from iron minerals, scale, and rusty tubular goods. Iron control agents isolate or chelate the iron and therefore prevent the formation of the iron precipitate. Nash-El-Din et al. [30] suggested the use of iron control agents, which are added to HCl solutions to prevent the precipitation of ferric hydroxide once the acid is spent. One way to prevent precipitation is to use reducing agents such as erythorbic acid. Another approach is to use the chelating agents mentioned above or citric acid [10] and nitrilotriacetic acid and its sodium salt [30]. Sometimes a non-corrosive chelating solvent applicable for dissolving carbonate scale is added to the CIF, such as the ammonium salts of EDTA (ethylenediaminetetraacetic acid), HEDTA (N-(hydroxyethyl)-ethylenediaminetriacetic acid), and DTPA (diethylenetriaminepentaacetic acid) – an amount from 0.1% to 15% may be present in the CIF [19],[27] and [117].

Wetting agents are employed to facilitate the penetration of the acid into the cracks and fissures in the scale, which helps remove the scale. They are known as pickling accelerators and usually do not have a corrosion inhibition effect [118]. Some hydrocarbons may form acid sludge in the presence of live or spent acid mixtures. Due to this reason, anti-sludge agents, which are surfactants, are added to the acidizing fluid to prevent sludge formation [24] and [119]. An anti-sludge agent is usually used in low concentration (no higher than 1.0 wt.%). Some anti-sludge agents can also act as non-emulsifiers.

Non-emulsifiers such as dodecylbenzylsulphonic acid (DDBSA) are employed to prevent the mixing of the acid and the extracted crude oil, therefore to prevent oil–acid emulsions.

Stabilizers are also added to the CIF to reduce the precipitation of the CIs on the rocks (some examples of stabilizers are given below by Walker [12]). Sometimes different dispersing agents are needed to disperse the solution better. Examples of dispersing agents are aromatic amines, aliphatic amines, and heterocyclic amines, such as aminophenol, aniline, chloroaniline, toluidine, diphenyl amine, picoline, alkyl pyridine, and n-octylamine [15] and [64].

Sometimes a viscoelastic agent is used to gel the system at intermediate pH levels [15], which eliminates the need for multiple stages. The acid treating fluid is initially at a low pH and the viscoelastic agent has a very low viscosity, which makes the acid treating fluid easy to pump and flow into the pores and channels of the formation. Upon acid reaction with the rock formation, the viscosity of the fluid increases due to the increase in the calcium ion content and pH, thus causing *in situ* gelling of the acid. The higher viscosity of the gelled viscoelastic agent temporarily blocks the wormholes and channels formed in the formation, allowing the acid to divert to other untreated areas. The viscosity of the gelled acid can be completely reduced by the introduction of a mutual solvent or by the produced hydrocarbons during flow-back.

Glycol and methanol are often added to flowing systems to decrease the corrosion activity of the aqueous solutions. It was also assumed that consequently a change in the CO_2 corrosion mechanism occurs [120].

The Effectiveness of Different Corrosion Inhibitor Formulations (CIFs)

As mentioned, the inhibitors frequently used for CIF design in acidizing procedures include acetylenic alcohols, -alkenylphenones, , -unsaturated aldehydes, quaternary amines, and derivatives of pyridinium and quinolinium salts. They are commonly formulated with solvents, surfactants, and intensifiers. Some examples are given below.

Beale and Kucera [121] tested different combinations of acetylenic alcohols (preferably those in Fig. 1) as CIFs for C1010 MS in HCl, H_2SO_4, sulphonic, phosphoric, and acetic acids at 93.3°C. These combinations allowed the use of smaller total amounts of the inhibitor. They claim that there is an advantage in mixing more than 2 compounds and that the most pronounced effect was observed when the acetylenic alcohols were present in substantially equal amounts. The preferred combinations comprise a lower molecular mass compound containing 3–6 carbon atoms and a higher molecular mass compound containing about 7–11 carbon atoms.

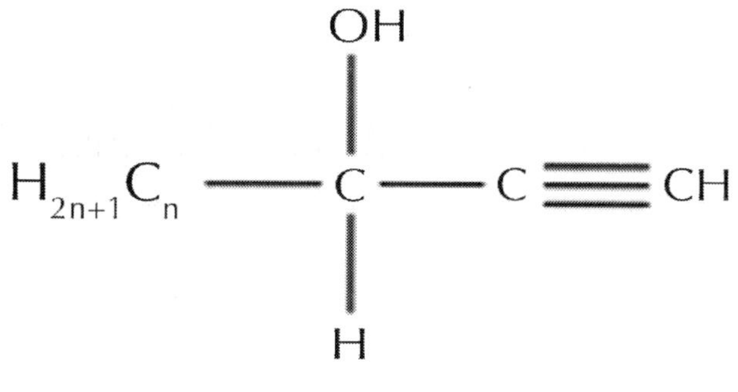

$$H_{2n+1}C_n \underset{\displaystyle H}{\overset{\displaystyle OH}{\mid}} C - C \equiv CH$$

Figure 1: Preferred structure of the acetylenic alcohols by Beale and Kucera [121], where n is an integer from 0 to about 8.

Keeney and Johnson [40] claim that a CIF consisting of nitrogen-containing compounds or acetylenic alcohol compounds or their mixtures, and CuI (at 25–25,000 mg/L by weight) is effective for ferrous materials corrosion protection in HCl, H_2SO_4, HF, acetic acid, and mixtures thereof at 65.5–232.2°C. Among the acetylenic alcohols, they suggested hexynol, dimethyl hexynol, dimethyl hexynediol, dimethyl hexynediol, dimethyl octynediol, methyl butynol, methyl pentynol, ethynyl cyclohexanol, 2-ethyl hexynol, phenyl butynol, and ditertiary acetylenic glycol, butynediol, 1-ethynylcyclohexanol, 3-methyl-1-nonyn-3-ol, 2-methyl-3-butyn-2-ol, 1-propyn-3-ol, 1-butyn-3-ol, 1-pentyn-3-ol, 1-heptyn-3-ol, l-octyn-3-ol, 1-nonyn-3-ol, 1-decyn-3-ol, and 1-(2,4,6-trimethyl-3-cyclohexenyl)-3-propyne-l-ol. Instead of acetylenic alcohols, acetylenic sulphide-type molecules may also be employed, with the general structure given in Fig. 2, such as dipropargyl sulphide, bis(1-methyl-2-propynyl) sulphide, and bis(2-ethynyl-2-propyl) sulphide. For nitrogen-containing compounds, they suggested amines such as mono-, di-, and tri-alkyl amines having 2–6 carbon atoms in each alkyl moiety, such as ethylamine, diethylamine, triethylamine, propylamine, dipropylamine, tripropylamine, mono-, di-, and tri-butylamine, mono-, di-, and tri-pentylamine, mono-, di-, and tri-hexylamine, and isomers of these, such as isopropylamine and tertiarybutylamine. They also suggested the six-membered N-heterocyclic amines, e.g. alkyl pyridines, having 1–5 nuclear alkyl substituents per pyridine moiety, with alkyl substituents having from 1 to 12 carbon atoms and preferably those having an average of six carbon atoms per pyridine moiety.

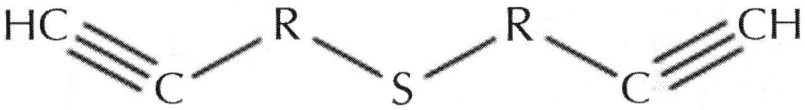

Figure 2: The structure of acetylenic sulphides suggested by Kenney and Johnson [40].

Growcock and Frenier [108] tested *trans*-cinnamaldehyde as a CI for API J55 steel in HCl solution at 65°C. They also tested the synergistic effect of *trans*-cinnamaldehyde with 3 surfactants, i.e. *N*-dodecylpyridinium bromide (DDPB), the adduct of trimethyl-1-heptanol with 7 mol of ethylene oxide (THEO), and Polystep A18 (commercial name), which is a sulphonate. The authors proposed that *trans*-cinnamaldehyde adsorbs onto protonated active sites to form a tenacious surface species, which serves as a primary barrier to mitigate corrosion. Moreover, they claimed that *trans*-cinnamaldehyde subsequently polymerizes on the surface and that the time-dependent polymerisation may be initially assisted by the surfactants. This was confirmed in later studies [122] and [123]. Subsequently, Growcock et al. [109] performed a similar study using different derivatives of cinnamaldehyde and showed that these compounds can act as effective CIs, especially when formulated with the above-mentioned surfactants. Growcock [60] also showed that the mixture of -alkenylphenone and DDPB effectively protects API J55 steel from corrosion in HCl solution at acid concentrations up to 28.3% and 95°C. In addition, Frenier et al. [27] also observed that this class of alkenylphenones is effective in 15–28% HCl at 65°C, which have the following structure: where in R_1 is an unsubstituted or inertly substituted aryl of 6–10 carbon atoms, and R_2 and R_3 are the same or different and each can be hydrogen, halogen, or an inertly substituted aliphatic of about 3–12 carbon atoms, and R_2 may also be an alkanol, ether, or unsubstituted or inertly substituted aryl of 6–10 carbon atoms, provided that the total number of carbon atoms in the alkenylphenone does not exceed 16. Inert substituents means that they do not have an effect on the corrosion inhibition of the corresponding unsubstituted alkenylphenone. The formulation containing alkenylphenone preferably includes a surfactant at concentrations up to 2%. They suggested surfactants such as THEO, DDPB, 4-decylated oxydibenzenesulphonate, and coco beta-amino propionate. Furthermore, Frenier et al. [61] demonstrated that octynol without surfactant protects J55 CS effectively at T up to 93°C. Moreover, they reported that CIF containing 2-benzoyl-allyl alcohol, 2-benzoyl-3-methoxy-1-propene, 2-benzoyl-1, 3-dimethoxy-propene, and 5-benzyol-1, 3-dioxane in combination with the surfactants DDPB and THEO protect J55 CS effectively at T up to 93°C.

Frenier [25] also observed that mixtures of alkenylphenones (Fig. 3) and N-substituted quinolonium salts (Fig. 4) are effective CIFs for iron and steel corrosion protection over a broad range of HCl concentrations and at T up to 200 C in HCl, HF, H_2SO_4, H_3PO_4, formic acid, acetic acid, citric acid, and their mixtures. R_4 in Fig. 4 is an alkyl group of about 4 to about 16 carbon atoms, or an alkylaryl of about 7 to about 20 carbons, and X is chlorine or bromine. The author preferred quinolinium salt composed of 1-(-naphthylmethyl)-quinolinium chloride. This CIF may also contain the same surfactant as used before [27] and an intensifier such as propionic, propiolic, formic, acetic, and chloroacetic acids or halide ions. Moreover, this CIF may also contain EDTA, an ammonium salt of EDTA, HEDTA, and DPTA chelating agents.

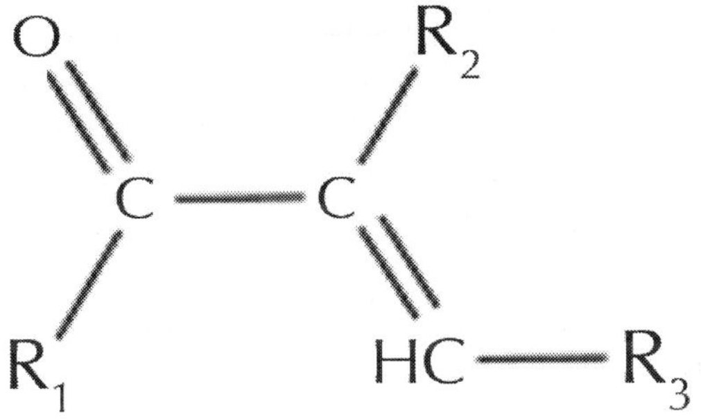

Figure 3: The structure of alkenylphenones proposed by Frenier et al. [27].

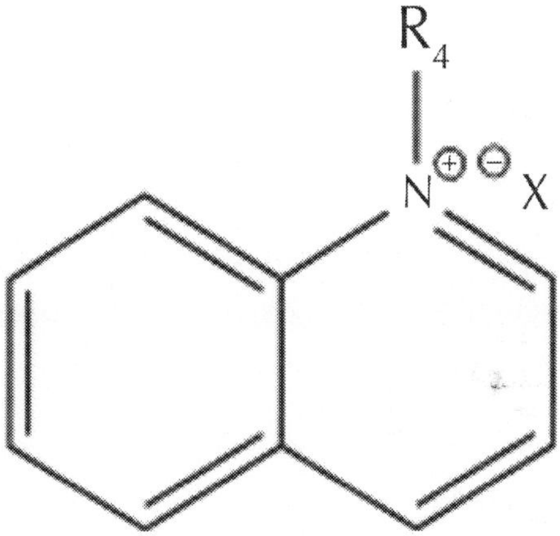

Figure 4: The structure of N-substituted quinolones proposed by Frenier [25].

Jasinski and Frenier [62] suggested a CIF designed for steel with Cr content higher than 9%, composed of phenyl ketone, phenyl ketone with a quaternary salt of a nitrogen-containing heterocyclic aromatic compound, or cinnamaldehyde (cinnamaldehyde can be employed substituted or unsubstituted) with a quaternary salt of a nitrogen-containing heterocyclic aromatic compound and an acid soluble metal from antimonium or bismuth (such as Bi_2O_3) salts. Moreover, HCOOH or its derivatives may be employed to even increase the performance of the CIF, especially when Sb ions are present, preferably from Sb_2O_3 and $SbCl_3$. This CIF was designed for HCl or a mixture of HCl/HF at temperatures above 121°C up to 246°C. The phenyl ketones may be C_{9-20} -alkenylphenones or hydroxyalkenylphenones and their mixtures. Among them, the authors suggested 2-benzoyl-3-hydroxy-1-propene, 2-benzoyl-3-methoxy-1-propene, and phenyl vinyl ketone. As a nitrogen-containing heterocyclic aromatic quaternary salt, a pyridinium, quinolinium, isoquinolinium, benzoazolinium, or benzothiazolinium salt may be used. They especially suggested N-cyclohexylpyridinium bromide, N-octylpyridinium bromide, N-nonylpyridinium bromide, N-decylpyridinium bromide, N-dodecylpyridinium bromide, N, N-dodecyldipyridinium dibromide, N-tetradecylpyridinium bromide, N-laurylpyridinium chloride, N-dodecylbenzylpyridinium chloride, N-dodecylquinolinium bromide quinolinium-(l-naphthylenemethyl) chloride, and N-naphthylmethyl quinolinium chloride. Of these, the authors prefer naphthylmethyl quinolinium chloride and dodecylpyridinium bromide. Finally, it was noted in the paper that the CIF combinations of the above-mentioned inhibitors with Cu_2Cl_2/ KI were more effective compared with combinations with Cu_2Cl_2/ HCOOH.

Fernier and Growcock [117] also proposed a CIF composed of an , -unsaturated aldehyde (Fig. 5) and a surfactant which is effective for ferrous materials as well as for Al, Zn, and Cu in aqueous acids such as HCl, HF, H_2SO_4, H_3PO_4, formic acid, acetic acid, citric acid, and their mixtures. This CIF is also effective for the above-mentioned materials in alkaline solutions and brines. R_1 in Fig. 5 represents a substituted or non-substituted saturated or unsaturated aliphatic hydrocarbon group containing from about 3 to about 12 carbon atoms with or without one or more non-interfering substituents, an aryl group (e.g. phenyl, benzyl, or the like), or an aryl group containing one or more non-interfering substituents. R_2 in Fig. 5 represents hydrogen, a saturated or unsaturated aliphatic hydrocarbon group containing from 1 to about 5 carbon atoms with or without one or more non-interfering substituents, an aryl group, or a substituted aryl group containing one or more non-interfering substituents. R_3 in Fig. 5 represents hydrogen, a saturated or unsaturated aliphatic hydrocarbon group containing from about 3 to about 12 carbon atoms with or without one or more non-interfering substituents, and an aryl group with or without one or more non-interfering substituents. The total number of carbon atoms in the substituents represented by R_1, R_2, and R_3 range from 1 to about 16, and preferably from about 5 to about 10. The non-

interfering substituents which may replace hydrogen on the - and
 -carbon atoms of the aldehydes in Fig. 5, or which are found in the
hydrocarbon substituents which replace hydrogen on these carbon
atoms, have no adverse effect on the corrosion inhibition and are,
e.g. lower alkyl (containing from 1 to about 4 carbon atoms), lower
alkoxy (containing from 1 to about 4 carbon atoms), halo, i.e. fluoro,
chloro, bromo, or iodo, hydroxyl, dialkylamino, cyano, thiocyano,
N,N-dialkylcarbamoylthio, and nitro substituents. The authors also
expanded on the use of surfactants that can be employed in CIFs. These
can be of the anionic, cationic, non-ionic, and amphoteric types.
Examples of surfactants in this reference were: alkylsulphates, such
as sodium alkyl sulphate, alkyl aryl sulphates, such as polypropylene
benzene sulphonates, and dialkyl sodium sulphosuccinates, such
as   dioctyl   sodium   sulphosuccinate,   N-cyclohexylpyridinium
bromide,   N-octylpyridinium   bromide,   N-nonylpyridinium   bromide,
N-decylpyridinium bromide, N-dodecylpyridinium bromide, N,N-
dodecyldipyridinium   dibromide,   N-tetradecylpyridinium   bromide,
N-laurylpyridinium   chloride,N-dodecylbenzylpyridinium   chloride,
N-dodecylquinolinium   bromide   quinolinium-(1-naphylenemethyl)
chloride, monochloromethylated and bischloromethylated pyridinium
halides,   ethoxylated   and   propoxylated   quaternary   ammonium
compounds, sulphated ethoxylates of alkyl phenols and primary
and   secondary   fatty   alcohols,   didodecyldimethylammonium
chloride,   hexadecylethyldimethylammonium   chloride,   2-hydroxy-
3-(2-undecylamidoethylamino)-propane-l-triethylammonium
hydroxide,   2-hydroxy-3-(2-heptadecylamidoethylamino)-
propane-1-triethylammonium   hydroxide,   2-hydroxy-3-(2-
heptadecylamidoethylamino)-propane-1-triethylammonium
hydroxide, primary amines, secondary amines, tertiary amines (e.g.
dodecyl dimethyl amine), ethoxylates of alkyl phenols, primary fatty
alcohols, secondary fatty alcohols, polyoxyethylenepolyoxypropylene
block copolymers, and coco- -aminopropionate.

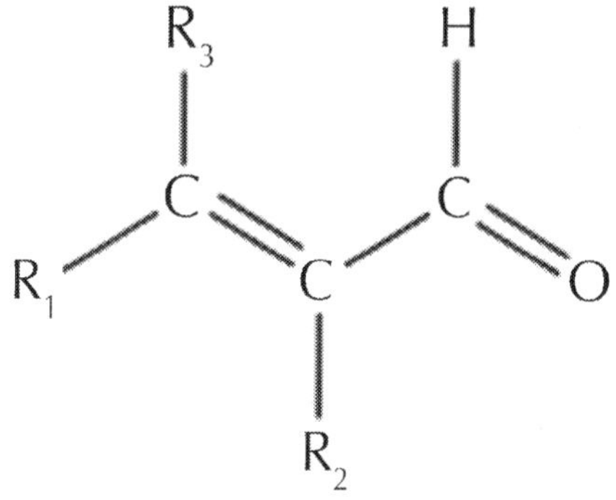

Figure 5: The structure of the α, β-unsaturated aldehydes proposed by Fernier and Growcock [117].

Gao et al. [124] showed that different α, β-unsaturated carbonyl compounds (cinnamaldehyde, benzalacetone, phenyl styryl ketone) formulated with propargyl alcohol act as very efficient CIFs for N80 steel in 20% HCl at 90 °C. The authors claim that the main reason for the high η at elevated temperatures is the polymerization and adsorption of these compounds on the steel surface. Moreover, Sastri [125] pointed out that commercial CIFs for use at high temperatures invariably contain acetylenic alcohols. The high η of propargyl alcohol is attributed to the iron complex catalysed formation of protective polymer films, which is favoured at high temperatures.

Williams et al. [16], [64] and [126] claimed that a CIF containing quaternary ammonium compound, metal ions, a highly polar aprotic solvent, and a surfactant is effective in mitigating corrosion of well construction steel (N-80, Cr 2205 [64] and J-55, P-105, Cr-9, Cr-13, Cr-2205, and Cr-2250 [16] and [126]) during the acidizing treatment with HCl, HF, formic acid, acetic acid, and/or their mixtures. The preferred quaternary ammonium compounds in this CIF are the following: alkyl pyridine-N-methyl chloride quaternary, alkyl pyridine-N-benzyl chloride quaternary, quinoline-N-methyl chloride quaternary, quinoline-N-benzyl chloride quaternary, quinoline-N-(chloro-benzyl chloride) quaternary, isoquinoline quaternaries, benzoquinoline quaternaries, chloromethyl naphthalene quaternaries, and chloromethyl naphthalene quinoline quaternaries. Among the metal compounds, they conclude that they should be present at concentrations of at least 0.08 wt.% (0.04 wt.% in the case of Sb-compound [16] and [126]), Sb- [16] and [126], Bi-, Ca-, and Cu(I)-compounds are the most preferable. Williams et al. [111] also suggested a mixture of at least two metal ions, where the first metal compound is selected from an antimony, bismuth, and cuprous compound and the second metal ion is selected from Ca, Al, Mg, Zn, and Zr ions. The suggested highly polar aprotic solvents are dimethyl formamide, dimethylsulphoxide, dimethylacetamide, 1-methyl-2-pyrrolidone, tetramethylene sulphone, and their mixtures. These solvents may be blended with (most preferably) dimethyl formamide. This CIF may also contain a dispersant such as an organic amine (including aromatic amines, aliphatic amines, and heterocyclic amines). Of these, the authors prefer aminophenol, aniline, chloroaniline, toluidine, diphenyl amine, picoline, alkyl pyridine, or n-octylamine. As a surfactant, the authors suggested ethoxylated alkyl phenols, ethoxylated aliphatic alcohols, polyethylene glycol esters of fatty, resin, and tall oil acids.

Furthermore, Williams et al. [14] described a CIF containing bismuth compound (0.4–1.4%), a quaternary ammonium compound (0.4–2.2%), and a surfactant (0.1–1.5%) to inhibit corrosion in well construction steels (such as N-80, J-55, P-105, Cr-9, Cr-13, Cr-2205, and Cr-2250) in HCl, HF, or their mixtures. The concentrations in brackets represent the most preferable content in the CIF. This CIF was designed to avoid a toxic combination of Sb-compounds and acetylenic alcohols [111]. The preferred quaternary ammonium compounds in this CIF are the same as described above for [64], except for chloromethyl naphthalene quinoline quaternaries. The authors suggested that the most preferable compounds are those containing a benzyl group. The quaternary compound:Bi ratio may be used in molar ratios of 1:1–5:1. The same surfactant as mentioned above is also suggested for use in this CIF [64].

Coffey et al. [107] described two CIFs and claim that they are effective for ferrous materials in hydroxyacetic, acetic, propionic, formic, HCl, HF, H_2SO_4, and H_3PO_4 acids and their mixtures, especially in the presence of H_2S. The first CIF includes a formaldehyde or paraformaldehyde (the latter is preferred), an acetophenone or its derivatives, a cyclohexylamine or its derivatives (e.g. 2-methyl cyclohexylamine or 2,4-dimethyl cyclohexylamine) and optionally, an aliphatic carboxylic, an acid such as octanoic acid, myristic acid, pelargonic acid, lauric acid, oleic acid, and tall oil. The second CIF includes an acetylenic alcohol or a mixture thereof (such as 1-propyn-

3-ol, l-butyn-3-ol, 1-pentyn-3-ol, 1-hexyn-3-ol, 1-heptyn-3-ol, l-octyn-3-ol, 1-nonyn-3-ol, 1-decyn-3-ol, or 1-octyn-4-ethyl-3-ol), an excess of formaldehyde, and optionally a surfactant (a non-ionic one is preferred, such as ethoxylated alkanols or ethoxylated alkyl phenols) and alcohols with 1–4 carbon atoms (they preferred isopropanol).

Walker [28] described a CIF for acidic solution in acidizing subterranean formations with ferrous metal well bores at T of 65.5–260 C. The acidic solutions that were described as mineral acids are HCl, or mixtures of HCl with HF, acetic acid, formic acid, or HF, H_2SO_4, formic acid, acetic acid, and their mixtures. The formulation was based on one or more acetylenic alcohols (5–35% of the amount of the formulation) with the structure given in Fig. 6, a quaternary ammonium compound, an aromatic hydrocarbon having high oil-wetting characteristics, and any antimony compound which is capable of activation by the other constituents of the CI. The structure of these acetylenic alcohol compounds is different than the one above in Fig. 1 proposed by Beale and Kucera [121]. Walker [28] proposed that the acetylenic alcohols employed having the general formula shown in Fig. 6, where R_1, R_2, and R_3 are hydrogen, alkyl, phenyl, substituted phenyl, or hydroxy-alkyl radicals. He suggested that preferably R_1 comprises hydrogen, R_2 comprises hydrogen, methyl, ethyl, or propyl radicals, and R_3 comprises an alkyl radical having the general formula C_nH_{2n}, where n is an integer from 1 to 10. He proposed acetylenic alcohols such as methyl butynol, methyl pentynol, hexynol, ethyl octynol, propargyl alcohol, benzylbutynol, and ethynylcyclohexanol. The most preferable selection among them was hexynol, propargyl alcohol, methyl butynol, and ethyl octynol. Among quaternary ammonium compounds, Walker suggested the same as Williams et al. [64] (see above), except quinolone-N-(chloro-benzyl chloride) quaternary and chloromethyl naphthalene quinoline quaternaries. As a hydrocarbon compound, which exhibits high oil-wetting characteristics, Walker suggested xylenes, saturated biphenyl-xylenes admixtures, HAN (heavy aromatic solvent), tetralene, tetrahydroquinoline, and tetrahydronaphthalene. The antimony compound, preferably at 0.7–40 mmol/L concentration, can be comprised of antimony trioxide, antimony pentoxide, antimony trichloride, antimony sulphide, antimony pentachloride, potassium antimony tartrate, antimony tartrate, antimony trifluoride, potassium pyroantimonate, antimony adducts of ethylene glycol, solutions containing ethylene glycol, water and the oxidized product of hydrogen peroxide, and antimony trioxide or any other trivalent antimony compound. Walker also suggested that this CIF can be dissolved in an alkanol solvent such as methyl, ethyl, propyl, isopropyl, butyl, pentyl, hexyl, heptyl, or octyl alcohol. This CIF can also comprise a non-ionic surfactant, which facilitates the dispersion of the CI in the acidic solution, such as ethoxylated oleate, tall oils, or ethoxylated fatty acids, preferably at volumes up to 20%. The CI could be generated *in situ* in the acidic solution, if so, then Walker suggested mixing all the constituents prior to addition of the antimony compound.

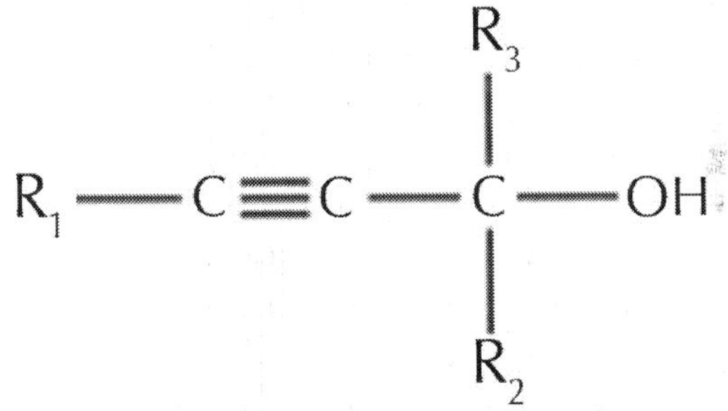

$$R_1 - C \equiv C - \underset{\underset{R_2}{|}}{\overset{\overset{R_3}{|}}{C}} - OH$$

Figure 6: The structure of the acetylenic alcohols proposed by Walker [28].

Additionally, Walker [12] used the same CIF as above [28], but with the addition of a stabilizer, which substantially prevents the precipitation of solubilized antimony-containing compounds from aqueous solutions and mitigates steel corrosion at T 65.5–260 C. The stabilizer can be one of the reactive fluoride-containing compounds, compounds having - or -hydroxy organic acid functional groups, or non-organic acid polyhydroxy compounds having 3–9 carbon atoms. The examples of the fluoride-containing stabilizers are HF, ammonium bifluoride, sodium fluoride, potassium fluoride, ammonium fluoride, transition metal fluorides, rare earth fluorides, and alkaline earth fluorides. The compounds having -hydroxy or -hydroxy organic acid functional groups are citric acid, citric acid salts, tartaric acid, tartaric acid salts, glycolic acid, glycolic acid salts, lactic acid, lactic acid salts, 3-hydroxyl propionic acid, 3-hydroxyl-butanoic acid, and 3, 4-dihydroxy-1, 6-hexanedioic acid. The suggested non-organic acid polyhydroxy compounds are sorbitol, glycerol, glucose, mannose, ribitol, erythritol, mannitol, perseitol, iditol, altritol, and xylitol. This stabilizer may be admixed with the acidic solution either before or after the addition of the antimony compound.

Walker [13] suggested a CIF which contains several components. Due to the numerous compounds possible for such a CIF design, it is recommended that readers themselves review the patent published. In general, the patent covers a CIF with the following criteria: (a) a compound having at least one reactive hydrogen atom and having no groups reactive under reaction conditions other than hydrogen, (b) a carbonyl compound having at least one hydrogen atom on the carbon atom adjacent to the carbonyl group, (c) an aldehyde, and (d) a fatty compound and an acid source which is admixed with a source of antimony ions. The main purpose was to avoid the usage of acetylenic alcohols.

Ali et al. [15] disclosed a treatment fluid for iron-containing materials comprising a mineral acid, a viscoelastic surfactant gelling agent, and a CI system containing at least one of the following: an alkenylphenones (Fig. 3) or α,β-unsaturated aldehyde (Fig. 5, cinnamaldehyde or its derivatives have been found to be particularly effective), an unsaturated ketone or unsaturated aldehyde other than the alkenylphenones and α,β-unsaturated aldehyde, a dispersing agent (such as an organic amine, also used before [64]), an extender (iodine) and an alcohol solvent. This CIF may also contain an intensifier mixture of CuI and CuCl. As the viscoelastic surfactant, they proposed erucylamidopropyl betaine surfactant. This CIF was designed to achieve a formic acid-free mixture, which it is claimed causes potential pipeline corrosion problems.

Baddini et al. [18] reported CIFs based on cinnamaldehyde, benzalacetone, and chalcone with propargyl alcohol, which were effective in reducing steel corrosion in 20% HCl at 90°C.

Barmatov et al. [9] studied different CIF combinations or individual compounds and their influence on the corrosion behaviour of HS80 and HS110 LCS (low carbon steel) in 14% HCl at 78°C. They reported that the of the surfactant DDPB as a CI increases sharply up to approximately 1 mmol/L concentration. For the cationic surfactant benzyldimethylhexadecylammonium chloride, they found that the increases with increased concentration and decreases with an elevation of temperature from 40°C to 78°C. Next, they determined the CR trend of some commercially available components as 2, 2'-biquinoline >

tripropargyl amine > 3-butyn-1-ol > 3-octyn-1-ol. For these compounds, it was reported that they all act as mixed-type inhibitors and that their increases with increased inhibitor concentration.

Similarly as before by Gao et al. [124] and Frenier et al. [61], Barmatov et al. [9] claimed that some acetylenic alcohols in combination with -alkenylphenones and α,β-unsaturated aldehydes may initiate surface polymerization, regarding which many authors believe that currently there is no alternative for the protection of oil well equipment during acid stimulation. Moreover, Barmatov et al. [9] pointed out, by investigating oil soluble 1-octyn-3-ol and 4-ethyl-1-octyn-3-ol and water soluble propargyl alcohol, that increases with increased chain length of the polymerizable acetylenic alcohols. Additionally, they stated that acetylenic alcohols in combination with quinolone-based quaternary ammonium compounds, a surfactant, and formic acid provide acceptable corrosion control at 104–177°C Finally, they reported that propargyl alcohol or 4-ethyl-1-octyn-3-ol in combination with DDPC show strong synergism.

Fischer and Parker [104] presented anhydrides derived from tall oil fatty acids (TOFA) and claim that they provide enhanced corrosion inhibition protection compared with traditional dimer/trimer acids (composed of 36 and 54 carbon atoms, respectively), due to the more tenacious film formed in the former case. These anhydrides are made by reacting maleic anhydride with the unsaturated fatty acids present in TOFA. The TOFA anhydride was neutralized with a fatty acid imidazoline. The authors reported that only one-seventh to one-tenth of the dosage of TOFA anhydride was required compared with that of an equivalent dimer/trimer-based active inhibitor to impart 90% corrosion protection in sweet and sour environments. Otherwise, we also noted that a commercially available CIF based on tall oil and established for HCl containing 1–5% methanol, 5–10% metyl formate, 10–30% formic acid, 30–60% ethoxylated tall oil (which contains palmitic, linoleic, and oleic acids).

Nasr-El-Din et al. [30] reported several different CIFs consisting of the mixtures of the following: (a) quaternary amines (15–40 wt.%), acetylenic alcohols (1–10 wt.%), prop-2-yn-1-ol (1–10 wt.%), naphthalene (1–5 wt.%), aliphatic hydrocarbons (30–60 wt.%), and propan-2-ol (5–10 wt.%), (b) a mixture containing quaternary ammonium salts (11–30 wt.%), benzyl chloride quaternary ammonium compound (11–30 wt.%), propargyl alcohol (1–10 wt.%), dimethyl formamide (1–10 wt.%), cuprous iodide (1–10 wt.%), ethoxylated nonylphenol (1–10 wt.%), and isopropanol solvent (10–30 wt.%), and (c) a mixture containing quaternary amines (10–20 wt.%), formamide (20–40 wt.%), acetylenic alcohols (5–10 wt.%), 2-propyn-1-ol (5–10 wt.%), ethoxylated nonylphenol (5–10 wt.%), pine oil (1–5 wt.%), and methanol and isopropanol solvent (20–40 wt.%). The authors claim that these CIFs are commonly used for HCl stimulation procedures.

Singh and Dey [116] studied the corrosion inhibition synergistic effect of propargyl alcohol with different inorganic cations and organic compounds for CRMS in 18% HCl at 33°C and 102°C. They reported a synergistic effect of propargyl alcohol with As^{3+}, Cr^{6+}, Cu^{2+}, Ni^{2+}, Hg^{2+}, and Sn^{2+}, which was dependent on both the propargyl alcohol and cation concentration. On the other hand, solutions containing all these ions alone except Sn^{2+} induced a higher CR at 33°C compared with the solution containing propargyl alcohol. A synergistic effect was also found when propargyl alcohol was formulated with phenol, formaldehyde, and sodium hypophosphide, but less when formulated with O-aminobenzoic acid.

Gao et al. [124] reported that cinnamaldehyde, benzalacetone, and chalcone, each formulated with propargyl alcohol, show a synergistic effect for reducing N80 steel CR in 20% HCl at 90°C. The authors claim that the main reason for the high corrosion   at elevated temperatures is the polymerization and adsorption of these compounds on the steel surface.

Kumar and Vishwanatham [127] tested three mixtures, i.e. formaldehyde:phenol, formaldehyde:o-cresol, and formaldehyde:p-cresol (all with a ratio of 1:2) as CIFs for N80 steel in 15% HCl at 25–115°C. They showed that all mixtures act as effective mixed-type inhibitors by predominantly reducing the rate of anodic reaction. Moreover, by using differential scanning calorimetry, the authors pointed out that the thermal stabilities of solutions containing cresol extend up to 200°C.

It should be noted that the usage of formaldehyde products is a problem due to their environmental unacceptability. As noted above, formaldehyde has been employed for various CIF designs in order to minimize hydrogen penetration into the steel [18]. Kumar and Vishwanatham [127], Baddini et al. [18], and Cardoso et al. [70]

presented a few such examples. However, Hill and Romijn [20] reported that formaldehyde is an animal carcinogen, therefore this limits its practical use.

Hill and Romijn [20] suggested the following chemistry for CIF design: (a) a mixture of phenyl vinyl ketones and acetylenic alcohols with the addition of potassium iodide and formic acid for J55, N80, and L80 steels at T up to 149 C, (b) formulation based on quaternary amine chemistry and cinnamaldehyde (also confirmed by Growcock et al. [109] with the addition of potassium iodide and formic acid, and nonylphenol ethoxylated or ethoxylated linear alcohol-based surfactant at T up to 121 C, (c) a mixture of phenyl vinyl ketones with potassium iodide and formic acid and a nonylphenol ethoxylated-based surfactant and toluene as a solvent and used for 13Cr steel, (d) quinolinium and pyridinium salts with antimony chloride [13] and [62], and (e) Mannich condensation product or quaternary salt with acetylenic alcohols, such as propargyl alcohol, 1-hexyl-3-ol (suggested also by Schmitt [66] and Sastri [128]) and 4-ethyl-1-octyn-3-ol, however the former two are very toxic by skin adsorption. Mannich bases are made by condensation of amines (mainly primary amines) with an aldehyde (mainly formaldehyde) and a ketone [118].

Sastri [128] and Schmitt [66] suggested the use of the following: (a) mixtures of N-containing compounds, acetylenic compounds, and surfactants, (b) condensation products of amines and aldehydes, (c) C_{12}–C_{18}–primary amines, cyclohexylamine, aniline, methylanilines, alkylpyridines, benzimidazole, and rosin amine condensed with formaldehyde, and (d) acetylenic inhibitors with Fe ions.

ENVIRONMENTAL CONCERNS IN CORROSION INHIBITION PROCESSES

As briefly mentioned, several components utilized in CIF have come under scrutiny regarding environmental health and safety (EH&S) issues. The emphasis is especially on finding environmentally acceptable acid CIs at elevated temperatures for acidizing environments. It is important to find non-toxic chemicals, with high biodegradability and reduced bioaccumulation. Environmental acceptability is usually assessed by the national regulations of a particular country. In particular, the North Sea is known for having the most stringent criteria regarding chemical qualifications. Most of the developed CIFs for the conventional acids no longer satisfy the OSPARCOM requirements, because their primary active ingredients may be harmful if discharged into the environment. OSPRAMCOM has the ultimate goal of replacing all environmentally hazardous chemical discharges by 2020 [110]. This presents a big problem for the existing CIFs mainly used for HCl and forces industry to replace or reformulate them. For example, CIFs containing acetylenic and antimony compounds present serious problems due to their high toxicity[64]. The goal of many research studies is to present reliable corrosion data to oilfield service companies in order to test CIF acceptability in large-scale operations in real field trials. It has to be emphasized that oilfield service companies are very interested in using safer and more environmentally acceptable alternatives than those currently employed, especially to satisfy OSPARCOM requirements [20].

In addition, particular countries are also now generating criteria for what can be classified as environmentally friendly. For example, WGK (German Wassergefährdungsklassen) stands for The German Water Hazard Class. The national German regulation, VwVwS (German *Verwaltungsvorschrift wassergefährdende Stoffe*), describes the water hazard classification such that all substances are either classified as non-hazardous to water or assigned to one of the three classes, WGK 1, WGK 2, and WGK 3, implying increasing water hazard. The lowest class, WGK 1, may seem relatively harmless and close to non-hazardous. However, if only one compound in the CIF design is in the WGK 2 or WGK 3 class and the others in WGK 1, the whole solution is classified as the higher WGK rating. The employment of more environmentally acceptable chemicals does not necessary mean that they will be less effective, but this is usually connected to the increased time needed to find a solution. However, most of the CIs that are still in use have hazardous effects on the environment [129].

ENVIRONMENTALLY FRIENDLY METHANESULPHONIC ACID (MSA)

Acids frequently represent a potential danger for drilling crews and the environment. For drilling-crews it is particularly important that fluids do not cause health problems, i.e. dermal toxicity, eye irritation, skin sensitisation, and mutagenicity. However, conventional acids frequently cause these problems. For example, HCl forms calcium chloride brine, which has been reported to cause skin injuries to workers in the oil industry [116]. Moreover, inhibitors for HCl are frequently effective only at high concentrations and are extremely toxic, causing handling and waste disposal problems, and producing toxic vapours under acidizing process conditions [61]. The dangers of using HF are also well known; e.g. the release of HF in a Nevada desert created a so-called death cloud. It was reported that 16 million Americans are potentially in a "kill zone" due to refineries using HF. Moreover, the very low LD50 value for chloroacetic acid (76 mg/kg for rats), which can also easily penetrate skin, is not safe to handle. The disadvantage of using formic and acetic acids is their volatility, which makes them difficult to handle.

Due to the above listed potential problems and disadvantages of using conventional acids, safer, environmentally more acceptable, and less corrosive alternatives are currently being sought. An inhibited MSA solution could be one of them. MSA is completely miscible in water and can be applied as a liquid over a wide temperature range. It is a strong organic acid ($pK_a = -1.9$) and has no tendency to either oxidize or reduce organic compounds. It has very low vapour pressure and a high boiling point, thus it is odour-free and evolves no dangerous volatiles. MSA salts are highly soluble, therefore it can be used in the well acidizing procedure. Moreover, MSA has low toxicity to aquatic life and is biodegradable within 28 days (which is a requirement for a chemical to be used in the North Sea [20]), with only CO_2 and sulphates being formed [130]. MSA is also present in the natural environment as part of the biogeochemical sulphur cycle, where atmospheric dimethyl sulphide arising from marine algae, cyanobacteria, and salt marsh plants is photochemically oxidized, leading to MSA formation. From the environmental perspective, MSA is usually described as a "green acid" [131] and is therefore environmentally much more acceptable compared with, e.g. HCl, HF, and chloroacetic acid. Moreover, stainless steel materials (e.g. duplex 22Cr and super-Cr-13) are usually used to combat H_2S and CO_2 corrosion. However, they are susceptible to corrosion in HCl solution [18]. On the other hand, we have recently shown that stainless steel materials are highly passivated in MSA solution [50]. Currently, extensive research by our research group is being carried out to evaluate the acceptability of MSA and the design of its CIFs for it to be used in the oilfield industry for the first time. MSA could be an alternative to the conventional acids currently employed due to its beneficial environmental properties and reduced hazard to the personnel involved.

RECOMMENDATIONS FOR TESTING CORROSION INHIBITOR ACCEPTABILITY

Before a CI or CIF is considered for field application, a suite of laboratory tests is performed for a particular application to evaluate its suitability [39], [132] and [133]. It is essential to perform laboratory tests under the same conditions as are in the pipe under actual conditions, for example at the same temperature and pressure, and to use the same coupon testing material as the pipe is made of. Usually, for experimental corrosion testing, specially designed glass equipment and autoclaves are used in order to simulate conditions in wells during the acidizing process. Sometimes in these tests, dissolved oxygen in the acid is not removed in order to simulate well stimulation [9]. The mixing order of the CIF components can also be important, e.g. Williams et al. [16] suggested using first the surfactant, then acetylenic alcohol, the solvent, the quaternary compound, and finally the metal intensifier. The references below list several key factors that can influence the results of acid corrosion testing. These include: the acid volume/sample area ratio [9],[24], [133] and [134], the surface preparation and cleaning [9], [25], [62], [132], [135], [136] and [137], the contact time and sample position [13], [19], [23], [67], [133] and [136], the emulsion stability test and temperature requirement for the liquid phase [19], the parasitic consumption [39], the temperature and pressure [19], [66] and [128], and the evaluation of pitting corrosion [9], [24], [132] and

[138]. As the majority of oilfield testing procedures for the evaluation of CI or CIF performance are done with WL tests, some of the important criteria are discussed below.

Acid Volume/Sample Area Ratio

Smith et al. [24] reported that the major discrepancy in corrosion inhibitor test results in reporting CR may occur by varying the ratio of the volume of the inhibited acid to the steel-coupon surface area. They observed a decrease in CR with increasing inhibited acid volume up to a ratio of (11.62 mL of acid solution)/(cm^2 of sample), where CR became constant. The authors explained that with an increase in the acid volume, the amount of the inhibitor increases, leading to better protection, because the steel area remains the same. They also suggested that the testing practise may require 25–150 mL of the acid solution per square inch of the sample (6.45 cm^2). It has to be pointed out that with a non-inhibited acid solution it may be just the opposite, because by decreasing the volume, saturation of the corrosion products in the acid solution may be achieved faster and the measured CR may be slower compared with the measurement in higher volumes. Moreover, the availability of oxygen may depend on the acid volume, which also influences CR [9]. The volume of acid per sample surface area should match the real operational procedure in the pipe. For example, a volume of solution per sample area ratio of 3.75 mL/cm^2 would simulate acidizing through a 15 cm diameter pipe. However, the latter example does not take into account that fresh acid is pumped into the system in the real acidizing situation. Also, under flow conditions, saturation of the corrosion products in the corrosion test does not occur [133]. Moreover, if the CR of the measured sample is high, consumption of the acid may be significant and its concentration changes, which does not simulate the real situation (freshly pumped acid). A drop from 28% to 19% HCl concentration after only 3 h of HS80 LCS immersion (closed system) was reported by Barmatov et al. [9]. Due to that reason, Barmatov et al. [9] claim that in laboratory tests, the results for CR higher than 0.243 kg/m^2 per test period may not be accurate and should be used as a rough estimate and calculated from WL may overestimate the inhibitors performance. On the other hand, the ASTM G31 [134] standard recommends a minimum solution volume of 0.4 mL per 1 mm^2 of the sample, which is quite high and makes the laboratory testing procedure impractical. However, Barmatov et al. [9] used 7 mL of acid solution per 1 cm^2 of sample.

In practice, the areas of the samples used for corrosion tests differ slightly, even though the samples look alike. Therefore, the precise area of each sample should be determined before the test and used for the calculation of the CR. The average value of replica CRs should be calculated from these numbers (the individual CR of a particular sample) and not according to the average mass-loss of the multiple samples, which could induce systematic error.

Surface Preparation and Cleaning

Smith et al. [24] claimed that metal sample surface preparation affects CR slightly. Moreover, Papavinasam et al. [132] reported that surface finish preparation (grinding and polishing) and slight differences in the metallurgy of the coupons have little effect on corrosion behaviour. On the other hand, Barmatov et al. [9] pointed out that the surface texture and roughness of LCS metallic samples may affect the CR and pitting formation, where surface stress plays an important role. This was especially pronounced for measurements at low CI concentrations (<0.14%). At higher CI concentrations, no significant effect of the surface texture and roughness on CR was observed. The authors compared 4 different preparation procedures: glass bead blasted (GBB) samples, pickled samples with HCl, ground samples with 240-grit SiC paper, and ground samples with 600- and 1200-grit SiC papers and afterwards polished with silica. For CI concentration of 0.01–0.05% in 14% HCl, the following CR order was reported: GBB > 240-grit ground samples > pickled samples > polished samples. Even though pickled samples had the highest surface roughness, their CR was slower compared with the samples with the GBB preparation procedure (the CR was also faster for the 240-grit ground samples). This was explained by the fact that the GBB procedure introduces stresses, plastic deformation, and microstrains, and changes in the heterogeneity of the surfaces. Moreover, in 28% HCl inhibited with 0.05% CI (which inhibitor was not reported), the GBB samples had a higher CR compared with the other 3 preparation procedures, for which the CR was similar. For the ground and polished samples, they did not observe any pits on the surface, whereas they reported pitting indexes (the definition is given below) of 3 and 7 for pickled and GBB samples, respectively.

Coupons are usually cut from metal sheet or from a real pipeline. In this manner, the cut edges can become sites of preferential corrosion attack, which is usually not experienced in a real pipe and the corrosion test thus would not simulate real field conditions [132]. To minimize an edge effect, it is preferred if the material is cut into sample coupons by using a water-cooled band saw to minimize changes in the material properties due to heat generated by the cutting procedure [9].

In general, to prepare a surface without deep scratches that could influence the CR and test results (especially electrochemical measurements), a circulating device is employed to grind the sample with, for example, up to 4000-grit SiC papers to ensure a uniform pattern of very shallow scratches. The grinding direction should be turned four times by 90°to minimize abrasion [43], [44], [45], [46], [47], [48], [49], [50], [51], [52], [53], [54], [55], [56] and [57]. Finally, in some cases polishing is subsequently best carried out according to the procedures provided by the company supplying the polishing material.

After grinding and polishing, the samples should be cleaned and degreased ultrasonically in a bath of acetone [25], methanol [135], or some other solution. However, it is important that this solution is not corrosive for the sample material. Barmatov et al. [9] suggested cleaning the samples in acetone prior the test and then drying them. After the preparation procedure and before the corrosion test, the samples must be stored in dry boxes (containing water adsorbent) to prevent atmospheric corrosion in the humid environment.

After corrosion tests, it is common practice to clean the specimens by rinsing with water or detergent solution, brushing with a fibre-bristle brush, and immersion in an ultrasound bath (containing special cleaning solutions). If corrosion products are still present on the surface, a procedure including dipping for 5–10 s one or more times in 10–15% HCl solution containing a CI (or possibly also Clarke solution), rinsing, and brushing is performed, which successfully removes corrosion products from the surface[136] and [137]. Some cleaning recommendations are provided in the ASTM G1 standard [137]. Barmatov et al. [9] suggested rinsing the samples in acetone and scrubbing them with soap and water to remove residual inhibitor film and corrosion deposits. Finally, they rinsed the samples in acetone before weighing. It has to be pointed out that the cleaning procedure after the corrosion test depends on the test conditions and what happened on the surface of the samples. Sometimes after the test the samples are "sticky" due to the inhibitor film and have to be cleaned in, e.g. acetone [62] or petroleum ether (however, this was not recommended in [136]). On the other hand, if a sample is covered with a lot of corrosion products, it is recommended to clean it in Clarke solution. After cleaning, the sample must be dried with inert gas (or by rinsing with anhydrous acetone or methanol) and immediately weighed. It is also important that during preparation, installation, and cleaning, the sample is handled with clean and dry gloves [136].

Contact Time

The API RP 13B-1 Standard Procedure for Field Testing Water-Based Drilling Fluids in Annex E explains that the contact time of the sample with the drilling fluid should be at least 40 h and up to 100 h (it is stated that 100 h is the normal exposure time for such test) [136]. The time needed to reach test temperature and to cool the test equipment is usually not included, but on the other hand, it increases the contact time of the sample with the acid solution. To decrease the influence of the cooling period, the equipment is commonly cooled down in ice, under a stream of water, or by employing an autoclave's cooling coils. It is important to pick the right testing period because the CR is usually fast at the beginning of the test and then decreases with an increase in the immersion time [67] and [136]. However, it could be just the opposite as well. Some authors suggest duplicating the well-treating environment in autoclave corrosion tests to ensure the worst case scenario for simulating real conditions [23]. HCl acid stimulation treatments frequently take 6–8 h [19] and that is most likely why Hill and Jones performed 8 h autoclave tests [19]. On the other hand, Walker [13] performed tests for a period of 4–48 h. Otherwise, it is practical and lately quite common to perform 24 h tests to obtain reliable CR results.

Evaluation of Pitting Corrosion

Localized attack that results in pits is the primary cause of corrosion-induced failures and thus a very important criterion in CIF design. A

WL experiment could show a small mass loss, but the pits formed can already be deep [24]. Extensive pitting corrosion of steel materials is very common in solutions containing high concentrations of chlorides. Extensive localized attack can lead to severe damage to the components. Special care should be taken in the case of tanks and pipes, where corrosion damage can cause the leakage of fluids or gases [46] and [47]. A good inhibitor must prevent significant pit development. Usually extensive pitting occurs when inhibitor concentrations are reduced to their absolute minimum [24]. One approach commonly used in the industry is visual observation and classification from 0 to 9 of pitting probability, according to the pitting index given in Table 3. Pitting represented by Ranks 1 through 4 is usually not considered serious [24]. On the other hand, Barmatov et al. [9] used a pitting index of 2 as a acceptability limit. It is also a common practice to measure the 3-D surface profile with, for example, a stylus or optical profilometer and express the pitting corrosion rate with the value of the maximum pit depth and the average of the ten deepest pits [138]. The use of this technique is still limited, because the scanned area is usually small and the analysis time is quite long.

Testing Techniques

At least three repetition measurements should be performed and the standard deviation should be calculated to present data with, for example, 95% confidence intervals. Outliers should be discarded according to statistical tests, such as the Grubbs statistical test [139].

Most of the literature quotes   and not CR (see Table 2). However, an inhibitor with calculated   of 90%, for example, could mean two things: (a) that the CI is very effective in preventing corrosion or (b) the reference CR for the non-inhibited solution was really high (e.g. 200 mm/y, which is not uncommon for CS in non-inhibited stimulation fluids), whereas the CR in the inhibited solution was slower, but still unacceptable (e.g. 20 mm/y). Moreover, in the literature the   of CIs is commonly reported to increase with increasing T (e.g. in [85]). However, this is usually due to an even faster CR for non-inhibited solution, whereas the CR also increases for the inhibited solution, but is slower compared with non-inhibited solution. The problem in reporting   instead of CR is even more emphasised when calculating   from electrochemical measurements (e.g. polarisation resistance measurements). Moreover, the Tafel extrapolation method for the determination of the corrosion current density from which the CR is calculated also causes problems (Eqs. (2) and (3)) [140] and [141]. This method was developed for kinetically controlled reactions. However, concentration polarisation, oxide formation, preferential dissolution of one alloy component, a mixed control process (where more than one anodic or cathodic reaction occurs simultaneously), and also other effects are frequent in corrosion measurements, which cause deviation from the original Tafel theory presented in Eq. (1) and consequently CR error.

$$j = j_{corr} \left(\exp\left(2.303\frac{E-E_{corr}}{\beta_a} \right) - \exp\left(-2.303\frac{E-E_{corr}}{\beta_c} \right) \right) \tag{1}$$

where j (current density) is the measured cell current, j_{corr} is the corrosion current density, E is the electrode potential, E_{corr} is the corrosion potential, and β_a and β_a are anodic and cathodic Beta Tafel coefficients, respectively (determined from measured curve slopes). Due to the problems listed above, it is recommended to report both and CR together, if possible. Calculation of and CR from the Tafel plot measurements are given in Eqs. (2) and (3) (j_{corr}^0 and j_{corr}^i represent j_{corr} measured in non-inhibited and inhibited solution, respectively).

$$\eta = 100\frac{j_{corr}^0 - j_{corr}^i}{j_{corr}^0} \tag{2}$$

$$Corrision\ rate = \frac{j_{corr}.K_1}{p.A.\sum_i^r (n_i w_i / A_i)} \tag{3}$$

where J_{corr} is in amperes (J – current), K_1 is the constant that defines the units for CR, is density (in g/cm³), A is the sample area (in cm²), n_i is the valence of the alloying element "i" in equivalent/mole, w_i is the mass fraction of the alloying element "i", A_i is the atomic mass of the element "i" in g/mol, and r is the number of elements in the alloy). It has to be emphasized that the CR calculation in Eq. (3) is only valid for uniform corrosion, but if localized corrosion occurs it dramatically underestimates the CR. To conclude, we believe that it is not recommended to calculate   and CR from Tafel plot measurements.

Papavinasam et al. [132] reported that WL measurements followed by the characterization of pits are by far the most reliable technique for monitoring the effect of CIs on uniform and pitting CRs in the oil and gas industry. Moreover, this technique is not affected by solution conductivity. The   and CR from WL measurements are calculated according to Eqs. (4) and (5).

$$\eta = 100 \frac{\Delta m^0 - \Delta m^i}{\Delta m^0} \qquad (4)$$

$$Corrision\ rate \left[\frac{mm}{y} \right] = \frac{K_2 \Delta m}{p.A.t} \qquad (5)$$

where K_2 is a conversion constant from cm h^{-1} to mm y^{-1} (87,600), m is the mass change (in grams, calculated from m before and after the corrosion test, m^0 and m^i represent m measured in non-inhibited and inhibited solution, respectively), and t is the immersion time (in h).

The polarisation resistance (R^0_p) measurement is also a convenient method for   determination, but less reproducible than WL measurement [132]. For the determination of the absolute CR in, for example, mm/y, the value of the anodic and cathodic Tafel slopes are needed, which causes uncertainty. It is better to use only R_p instead of calculated CR values, and compare CI performance on a relative basis instead. Moreover, it has been reported that error-producing complications when using R_p measurements include the oxidation of electroactive species besides the corroding metal, a change in the corrosion potential during R_p measurement, the use of a potential scan over too large a potential interval, the use of an excessively fast potential sweep rate, insufficient stabilization of the electrode before the measurement, uncompensated resistance (the biggest contribution is usually due to a non- or low-conducting medium), the presence of adsorbed intermediates, non-uniform current and potential distributions, and especially, as already mentioned above, incorrect values of the assumed or determined Tafel slopes [132] and [142]. It has to be emphasized that R_p measurement is a non-destructive method that can measure uniform corrosion behaviour over long time intervals, but it does not provide information on localized corrosion. calculated with R_p measurements is expressed as (R_p^0 and R_p^i represent R_p measured in non-inhibited and inhibited solution, respectively):

$$\eta = 100 \frac{R_p^i - R_p^0}{R_p^i}$$

(6)

The electrochemical impedance spectroscopy (EIS) technique was less recommended by Papavinasam et al. [132] in oil and gas pipeline corrosion inhibition monitoring due to poor reproducibility, long measurement time, and the need to develop a physical model (an equivalent electrical circuit). On the other hand, Lotz et al. [143] claim that EIS is a powerful technique for accessing *in situ* CR.

The electrochemical noise (EN) technique was highly recommended by Papavinasam et al. [132], however they reported that it is still in the developmental stage. The authors claim that no satisfactory method has yet been developed for the presentation and interpretation of EN data.

Papavinasam et al. [132] also reported that hydrogen permeation measurement does not correlate with real corrosion in oil and gas pipelines.

To conclude, an advantage of electrochemical over WL measurements is in obtaining other information beside CR and values, such as the following: which reaction of the corrosion couple is primarily inhibited, pitting and crevice corrosion susceptibility, repassivation ability after pitting formation, the passivation property of the material, the thermodynamics of redox processes, the kinetics of the electron transfer, and representation of the properties of the surface structure and its corrosion phenomena by equivalent electrical circuits.

CONCLUSIONS

This review summarizes the corrosion inhibition of lower-grade steels in acidic media. The focus herein was on HCl solutions and the elevated temperatures usually encountered in the well acidizing procedure. Lower-grade steel materials are the most commonly used construction materials for oil and gas wells due to their low cost and high performance. During the acidizing procedure these steel materials are under very corrosive conditions and need to be inhibited by means of an appropriate corrosion inhibitor. Numerous compounds were presented which are used as corrosion inhibitors for steel materials in acid solutions. This review should also prove useful for other fields of corrosion inhibitor research where conditions are not so severe, e.g. acid pickling, industrial cleaning, and acid descaling.

However, at elevated temperatures individual components alone are not effective in the well acidizing procedure and they often need to be formulated with appropriate surfactants, solvents, and intensifiers to protect metal in acidizing environments. Other components may also be present is such corrosion inhibitor formulations. In this work, various corrosion inhibitor formulations have been presented which are effective at elevated temperatures. It has been shown that acetylenic alcohols are the most widely used active components as corrosion inhibitors in formulation design, now for more than five decades.

This review also describes the potential danger of conventional acids commonly employed in the acidizing procedure and the use of methanesulphonic acid as an alternative acid solution for acidizing, due to its beneficial properties. Moreover, most of the developed corrosion inhibitors or their formulations no longer meet the Oslo Paris Commission requirements because their primary active ingredients may be harmful if discharged into the environment. Current research is thus focused on developing alternatives to those currently employed. Finally, some recommended references are reported herein that refer to important criteria in acidizing testing.

REFERENCES

1.   Z. Panossian, N.L.d. Almeida, R.M.F.d. Sousa, G.d.S. Pimenta, L.B.S. Marques Corrosion of carbon steel pipes and tanks by concentrated sulfuric acid: a review Corros. Sci., 58 (2012), pp. 1–11

2.   M.M. Osman, M.N. Shalaby Some ethoxylated fatty acids as corrosion inhibitors for low carbon steel in formation water Mater. Chem. Phys., 77 (2003), pp. 261–269

3.   P.C. Okafor, X. Liu, Y.G. Zheng Corrosion inhibition of mild steel by ethylamino imidazoline derivative in CO_2-saturated solution Corros. Sci., 51 (2009), pp. 761–768

4.   S. Nešić, W. Sun 2.25 – Corrosion in acid gas solutions, A.R. Tony (Ed.), Shreir's Corrosion, Elsevier, Oxford (2010), pp. 1270–1298

5.   V. Garcia-Arriaga, J. Alvarez-Ramirez, M. Amaya, E. Sosa H_2S and O_2 influence on the corrosion of carbon steel immersed in a solution containing 3 M diethanolamine Corros. Sci., 52 (2010), pp. 2268–2279

6.   S. Ghareba, S. Omanovic Interaction of 12-aminododecanoic acid with a carbon steel surface: towards the development of 'green' corrosion inhibitors Corros. Sci., 52 (2010), pp. 2104–2113

7.   D. Martínez, R. Gonzalez, K. Montemayor, A. Juarez-Hernandez, G. Fajardo, M.A.L. Hernandez-Rodriguez Amine type inhibitor effect on corrosion–erosion wear in oil gas pipes Wear, 267 (2009), pp. 255–258

8.   S.M. Wilhelm Galvanic corrosion in oil and gas production: Part 1—laboratory studies Corrosion, 48 (1992), pp. 691–703

9.   E. Barmatov, J. Geddes, T. Hughes, M. Nagl, Research on corrosion inhibitors for acid stimulation, in: NACE, 2012, pp. C2012–0001573.

10. S. Huizinga, W.E. Like Corrosion behavior of 13% chromium steel in acid stimulations Corrosion, 50 (1994), pp. 555–566

11. M.A. Quraishi, D. Jamal Technical note: CAHMT—a new and eco-friendly acidizing corrosion inhibitor Corrosion, 56 (2000), pp. 983–985

12. M.L. Walker, Method and Composition for Acidizing Subterranean Formations, in: US Patent 4,552,672, Halliburton Company, Duncan, Okla, 1985.

13. M.L. Walker, Method and Composition for Acidizing Subterranean Formations, in: US Patent 5,366,643, Halliburton Company, Duncan, Okla, 1994.

14. D.A. Williams, P.K. Holifield, J.R. Looney, L.A. McDougall, Method of Inhibiting Corrosion in Acidizing Wells, in: US Patent 5,200,096, Exxon Chemicals Patents, Inc., Linden, N.J., 1993.

15. S. Ali, J.S. Reyes, M.M. Samuel, F.M. Auzerais, Self-Diverting Acid Treatment with Formic Acid-Free Corrosion Inhibitor, in: US Patent 2010/0056405 A1, Schlumberger Technology Corporation, 2010.

16. D.A. Williams, P.K. Holifield, J.R. Looney, L.A. McDougall, Inhibited Acid System for Acidizing Wells, in: US Patent 5,209,859, Exxon Chemical Patents, Inc., Linden, N.J., 1993.

17. M. Yadav, D. Behera, U. Sharma, Nontoxic corrosion inhibitors for N80 steel in hydrochloric acid, Arab. J. Chem. http://dx.doi.org/10.1016/j.arabjc.2012.03.011.

18. A.L.d.Q. Baddini, S.P. Cardoso, E. Hollauer, J.A.d.C.P. Gomes Statistical analysis of a corrosion inhibitor family on three steel surfaces (duplex, super-13 and carbon) in hydrochloric acid solutions Electrochim. Acta, 53 (2007), pp. 434–446

19. D.G. Hill, A. Jones An engineered approach to corrosion control during matrix acidizing of HTHP sour carbonate reservoir Corrosion (2003) Paper No. 03121

20. D.G. Hill, H. Romijn Reduction of risk to the marine environment from oilfield chemicals: environmentally improved acid corrosion inhibition for well stimulation Corrosion (2000) Paper No. 00342

21. M. Quraishi, D. Jamal Fatty acid triazoles: novel corrosion inhibitors for oil well steel (N-80) and mild steel J. Am. Oil Chem. Soc., 77 (2000), pp. 1107–1111

22. M.A. Quraishi, N. Sardar, H. Ali A study of some new acidizing inhibitors on corrosion of N-80 alloy in 15% boiling hydrochloric acid Corrosion, 58 (2002), pp. 317–321

23. D. Brondel, R. Edwards, A. Hayman, D. Hill, S. Mehta, T. Semerad Corrosion in the oil industry Oilfield Rev., 6 (1994), pp. 4–18

24. C.F. Smith, F.E. Dollarhide, N.B. Byth Acid corrosion inhibitor: are we getting what we need? J. Petrol. Technol., 30 (1978), pp. 737–746

25. W.W. Frenier, Process and Composition for Inhibiting High-Temperature Iron and Steel Corrosion, in: US Patent 5,096,618, Dowell Schlumberger Incorporated, Tulsa, Okla, 1992.

26. M.A. Migahed, I.F. Nassar Corrosion inhibition of Tubing steel during acidization of oil and gas wells Electrochim. Acta, 53 (2008), pp. 2877–2882

27. W.W. Frenier, F. Growcock, V.R. Lopp, B. Dixon, Process and Composition for Inhibiting Iron and Steel Corrosion, in: US Patent 5,013,483, Dowell Schlumberger Incorporated, Tulsa, Okla, 1990.

28. M.L. Walker, Method and Composition for Acidizing Subterranean Formations, in: US Patent 4,498,997, Halliburton Company, Duncan, Okla, 1985.

29. D. Jayaperumal Effects of alcohol-based inhibitors on corrosion of mild steel in hydrochloric acid Mater. Chem. Phys., 119 (2010), pp. 478–484

30. H.A. Nasr-El-Din, A.M. Al-Othman, K.C. Taylor, A.H. Al-Ghamdi Surface tension of HCl-based stimulation fluids at high temperatures J. Petrol. Sci. Eng., 43 (2004), pp. 57–73

31. R. Wang, S. Luo Grooving corrosion of electric-resistance-welded oil well casing of J55 steel Corros. Sci., 68 (2013), pp. 119–127

32. D. Liu, Y.B. Qiu, Y. Tomoe, K. Bando, X.P. Guo Interaction of inhibitors with corrosion scale formed on N80 steel in CO_2-saturated NaCl solution Mater. Corros., 62 (2011), pp. 1153–1158

33. G.E. Badr The role of some thiosemicarbazide derivatives as corrosion inhibitors for C-steel in acidic media Corros. Sci., 51 (2009), pp. 2529–2536

34. W. Durnie, R. De Marco, A. Jefferson, B. Kinsella Development of a structure–activity relationship for oil field corrosion inhibitors J. Electrochem. Soc., 146 (1999), pp. 1751–1756

35. G. Zhang, C. Chen, M. Lu, C. Chai, Y. Wu Evaluation of inhibition efficiency of an imidazoline derivative in CO_2-containing aqueous solution Mater. Chem. Phys., 105 (2007), pp. 331–340

36. S. Vishwanatham, N. Haldar Furfuryl alcohol as corrosion inhibitor for N80 steel in hydrochloric acid Corros. Sci., 50 (2008), pp. 2999–3004

37. S.D. Zhu, A.Q. Fu, J. Miao, Z.F. Yin, G.S. Zhou, J.F. Wei Corrosion of N80 carbon steel in oil field formation water containing CO_2 in the absence and presence of acetic acid Corros. Sci., 53 (2011), pp. 3156–3165

38. A. Torres-Islas, S. Serna, J. Uruchurtu, B. Campillo, J. González-Rodríguez Corrosion inhibition efficiency study in a microalloyed steel for sour service at 50 C J. Appl. Electrochem., 40 (2010), pp. 1483–1491

39. D.I. Horsup, J.C. Clark, B.P. Binks, P.D.I. Fletcher, J.T. Hicks The fate of oilfield corrosion inhibitors in multiphase systems Corrosion, 66 (2010) 036001-036001-036014

40. B.R. Keeney, J.W. Johnson, Inhibited Treating Acid, in: US Patent 3,773,465, Halliburton Company, Duncan, Okla, 1973.

41. R.F. Monroe, C.H. Kucera, B.D. Oakes, N.G. Johnston, Compositions For Inhibiting Corrosion, in: US Patent 3,077,454, The Dow Chemical Company, 1963.

42. M. Finšgar, I. Milošev Inhibition of copper corrosion by 1, 2, 3-benzotriazole: a review Corros. Sci., 52 (2010), pp. 2737–2749

43. M. Finšgar Galvanic series of different stainless steels and copper- and aluminium-based materials in acid solutions Corros. Sci., 68 (2013), pp. 51–56

44. M. Finšgar 2-Mercaptobenzimidazole as a copper corrosion inhibitor: Part I. Long-term immersion, 3D-profilometry, and electrochemistry Corros. Sci., 72 (2013), pp. 82–89

45. M. Finšgar 2-Mercaptobenzimidazole as a copper corrosion inhibitor: Part II. Surface analysis using X-ray photoelectron spectroscopy Corros. Sci., 72 (2013), pp. 90–98

46. M. Finšgar, S. Fassbender, S. Hirth, I. Milošev Electrochemical and XPS study of polyethyleneimines of different molecular sizes as corrosion inhibitors for AISI 430 stainless steel in near-neutral chloride media Mater. Chem. Phys., 116 (2009), pp. 198–206

47. M. Finšgar, S. Fassbender, F. Nicolini, I. Milošev Polyethyleneimine as a corrosion inhibitor for ASTM 420 stainless steel in near-neutral saline media Corros. Sci., 51 (2009), pp. 525–533

48. M. Finšgar, J. Kovač, I. Milošev Surface analysis of 1-hydroxybenzotriazole and benzotriazole adsorbed on Cu by X-ray photoelectron spectroscopy J. Electrochem. Soc., 157 (2010), pp. C52–C60

49. M. Finšgar, A. Lesar, A. Kokalj, I. Milošev A comparative electrochemical and quantum chemical calculation study of BTAH and BTAOH as copper corrosion inhibitors in near neutral chloride solution Electrochim. Acta, 53 (2008), pp. 8287–8297

50. M. Finšgar, I. Milošev Corrosion behaviour of stainless steels in aqueous solutions of methanesulfonic acid Corros. Sci., 52 (2010), pp. 2430–2438

51. M. Finšgar, I. Milošev Corrosion study of copper in the presence of benzotriazole and its hydroxy derivative Mater. Corros.-Werkstoffe Korros., 62 (2011), pp. 956–966

52. M. Finšgar, I. Milošev, B. Pihlar Inhibition of copper corrosion studied by electrochemical and EQCN techniques Acta Chim. Slov., 54 (2007), pp. 591–597

53. M. Finšgar, S. Peljhan, A. Kokalj, J. Kovač, I. Milošev Determination of the Cu_2O thickness on BTAH-inhibited copper by reconstruction of auger electron spectra J. Electrochem. Soc., 157 (2010), pp. C295–C301′

54. A. Kokalj, N. Kovačević, S. Peljhan, M. Finšgar, A. Lesar, I. Milošev Triazole, benzotriazole, and naphthotriazole as copper corrosion inhibitors: I. Molecular electronic and adsorption properties ChemPhysChem, 12 (2011), pp. 3547–3555

55. Kokalj, S. Peljhan, M. Finšgar, I. Milošev What determines the inhibition effectiveness of ATA, BTAH, and BTAOH corrosion inhibitors on copper? J. Am. Chem. Soc., 132 (2010), pp. 16657–16668

56. M. Finšgar, D. Kek Merl, 2-mercaptobenzoxazole as a copper corrosion inhibitor in chloride solution: electrochemistry, 3D-profilometry, and XPS surface analysis Corros. Sci., 80 (2014), pp. 82–95

57. M. Finšgar EQCM and XPS analysis of 1, 2, 4-triazole and 3-amino-1, 2, 4-triazole as copper corrosion inhibitors in chloride solution Corros. Sci., 77 (2013), pp. 350–359

58. M. Finšgar, D. Kek Merl, an electrochemical, long-term immersion, and XPS study of 2-mercaptobenzothiazole as a copper corrosion inhibitor in chloride solution Corros. Sci., 83 (2014), pp. 164–175

59. Y.P. Khodyrev, E.S. Batyeva, E.K. Badeeva, E.V. Platova, L. Tiwari, O.G. Sinyashin The inhibition action of ammonium salts of O, O′-dialkyldithiophosphoric acid on carbon dioxide corrosion of mild steel Corros. Sci., 53 (2011), pp. 976–983

60. F.B. Growcock Corrosion kinetics of J55 steel in hydrochloric acid inhibited with benzoyl allyl alcohol Corrosion, 45 (1989), pp. 393–401

61. W.W. Frenier, F.B. Growcock, V.R. Lopp α-Alkenylphenones—a new class of acid corrosion inhibitors Corrosion, 44 (1988), pp. 590–598

62. R.J. Jasinski, W.W. Frenier, Process and Composition for Protecting Chrome Steel, in: US Patent 5,120,471, Dowell Schlumberger Incorporated, Tulsa, Okla, 1992, p. 7.

63. M.A. Quraishi, R. Sardar Dithiazolidines—a new class of heterocyclic inhibitors for prevention of mild steel corrosion in hydrochloric acid solution Corrosion, 58 (2002), pp. 103–107

64. D.A. Williams, P.K. Holifield, J.R. Looney, L.A. McDougall, Corrosion Inhibitor and Method of Use, in: US Patent 5,002,673, Exxon Chemical Patents Inc., Linden, N.J., 1991.

65. M.A. Quraishi, D. Jamal Corrosion inhibition of N-80 steel and mild steel in 15% boiling hydrochloric acid by a triazole compound—SAHMT Mater. Chem. Phys., 68 (2001), pp. 283–287

66. G. Schmitt Application of inhibitors for acid media Br. Corros. J., 19 (1984), pp. 165–176

67. M.A. Quraishi, D. Jamal Dianils: new and effective corrosion inhibitors for oil-well steel (N-80) and mild steel in boiling hydrochloric acid Corrosion, 56 (2000), pp. 156–160

68. B.I. Ita, O.E. Offiong The study of the inhibitory properties of benzoin, benzil, benzoin-(4-phenylthiosemicarbazone) and benzil-(4-phenylthiosemicarbazone) on the corrosion of mild steel in hydrochloric acid Mater. Chem. Phys., 70 (2001), pp. 330–335

69. E.A. Flores, O. Olivares, N.V. Likhanova, M.A. Domínguez-Aguilar, N. Nava, D. Guzman-Lucero, M. Corrales Sodium phthalamates as corrosion inhibitors for carbon steel in aqueous hydrochloric acid solution Corros. Sci., 53 (2011), pp. 3899–3913

70. S.P. Cardoso, J.A.C.P. Gomes, L.E.P. Borges, E. Hollauer Predictive QSPR analysis of corrosion Inhibitors for super 13% Cr steel in Hydrochloric acid Braz. J. Chem. Eng., 24 (2007), pp. 547–559

71. K. Babic-Samardzija, C. Lupu, N. Hackerman, A.R. Barron Inhibitive properties, adsorption and surface study of butyn-1-ol and pentyn-1-ol alcohols as corrosion inhibitors for iron in HCl J. Mater. Chem., 15 (2005), pp. 1908–1916

72. A. Popova, E. Sokolova, S. Raicheva, M. Christov AC and DC study of the temperature effect on mild steel corrosion in acid media in the presence of benzimidazole derivatives Corros. Sci., 45 (2003), pp. 33–58

73. Popova, M. Christov, S. Raicheva, E. Sokolova Adsorption and inhibitive properties of benzimidazole derivatives in acid mild steel corrosion Corros. Sci., 46 (2004), pp. 1333–1350

74. Popova, M. Christov, A. Zwetanova Effect of the molecular structure on the inhibitor properties of azoles on mild steel corrosion in 1 M hydrochloric acid Corros. Sci., 49 (2007), pp. 2131–2143

75. J. Aljourani, K. Raeissi, M.A. Golozar Benzimidazole and its derivatives as corrosion inhibitors for mild steel in 1 M HCl solution Corros. Sci., 51 (2009), pp. 1836–1843

76. K.F. Khaled The inhibition of benzimidazole derivatives on corrosion of iron in 1 M HCl solutions Electrochim. Acta, 48 (2003), pp. 2493–2503

77. K.F. Khaled, N. Hackerman Investigation of the inhibitive effect of ortho-substituted anilines on corrosion of iron in 1 M HCl solutions Electrochim. Acta, 48 (2003), pp. 2715–2723

78. K.F. Khaled, K. Babić-Samardžija, N. Hackerman Piperidines as corrosion inhibitors for iron in hydrochloric acid J. Appl. Electrochem., 34 (2004), pp. 697–704

79. J. Cruz, T. Pandiyan, E. García-Ochoa A new inhibitor for mild carbon steel: electrochemical and DFT studies J. Electroanal. Chem., 583 (2005), pp. 8–16

80. Z. Ait Chikh, D. Chebabe, A. Dermaj, N. Hajjaji, A. Srhiri, M.F. Montemor, M.G.S. Ferreira, A.C. Bastos Electrochemical and analytical study of corrosion inhibition on carbon steel in HCl medium by 1,12-bis(1,2,4-triazolyl)dodecane Corros. Sci., 47 (2005), pp. 447–459

81. O.K. Abiola Adsorption of 3-(4-amino-2-methyl-5-pyrimidyl methyl)-4-methyl thiazolium chloride on mild steel Corros. Sci., 48 (2006), pp. 3078–3090

82. M. Elachouri, M.S. Hajji, S. Kertit, E.M. Essassi, M. Salem, R. Coudert Some surfactants in the series of 2-(alkyldimethylammonio) alkanol bromides as inhibitors of the corrosion of iron in acid chloride solution Corros. Sci., 37 (1995), pp. 381–389

83. L. Tang, X. Li, L. Li, Q. Qu, G. Mu, G. Liu The effect of 1-(2-pyridylazo)-2-naphthol on the corrosion of cold rolled steel in acid media: Part 1: inhibitive action in 1.0 M hydrochloric acid Mater. Chem. Phys., 94 (2005), pp. 353–359

84. J. Cruz, R. Martı́nez, J. Genesca, E. Garcı́a-Ochoa Experimental and theoretical study of 1-(2-ethylamino)-2-methylimidazoline as an inhibitor of carbon steel corrosion in acid media J. Electroanal. Chem., 566 (2004), pp. 111–121

85. B.I. Ita, O.E. Offiong The inhibition of mild steel corrosion in hydrochloric acid by 2,2′-pyridil and α-pyridoin Mater. Chem. Phys., 51 (1997), pp. 203–210

86. S.S. Abd El Rehim, M.A.M. Ibrahim, K.F. Khalid The inhibition of 4-(2′-amino-5′-methylphenylazo) antipyrine on corrosion of mild steel in HCl solution Mater. Chem. Phys., 70 (2001), pp. 268–273

87. M.A. Quraishi, D. Jamal Corrosion inhibition by fatty acid oxadiazoles for oil well steel (N-80) and mild steel Mater. Chem. Phys., 71 (2001), pp. 202–205

88. M.A. Quraishi, D. Jamal, R.N. Singh Inhibition of mild steel corrosion in the presence of fatty acid thiosemicarbazides Corrosion, 58 (2002), pp. 201–207

89. A. Yıldırım, M. Çetin Synthesis and evaluation of new long alkyl side chain acetamide, isoxazolidine and isoxazoline derivatives as corrosion inhibitors Corros. Sci., 50 (2008), pp. 155–165

90. H.-L. Wang, R.-B. Liu, J. Xin Inhibiting effects of some mercapto-triazole derivatives on the corrosion of mild steel in 1.0 M HCl medium Corros. Sci., 46 (2004), pp. 2455–2466

91. S.A. Ali, M.T. Saeed, S.U. Rahman The isoxazolidines: a new class of corrosion inhibitors of mild steel in acidic medium Corros. Sci., 45 (2003), pp. 253–266

92. Mernari, H. El Attari, M. Traisnel, F. Bentiss, M. Lagrenee Inhibiting effects of 3,5-bis(n-pyridyl)-4-amino-1,2,4-triazoles on the corrosion for mild steel in 1 M HCl medium Corros. Sci., 40 (1998), pp. 391–399

93. F. Bentiss, M. Traisnel, H. Vezin, M. Lagrenée Linear resistance model of the inhibition mechanism of steel in HCl by triazole and oxadiazole derivatives: structure–activity correlations Corros. Sci., 45 (2003), pp. 371–380

94. F. Bentiss, F. Gassama, D. Barbry, L. Gengembre, H. Vezin, M. Lagrenée, M. Traisnel Enhanced corrosion resistance of mild steel in molar hydrochloric acid solution by 1,4-bis(2-pyridyl)-5H-pyridazino[4,5-b]indole: electrochemical, theoretical and XPS studies Appl. Surf. Sci., 252 (2006), pp. 2684–2691

95. F. Bentiss, M. Lebrini, M. Lagrenée Thermodynamic characterization of metal dissolution and inhibitor adsorption processes in mild steel/2,5-bis(n-thienyl)-1,3,4-thiadiazoles/hydrochloric acid system Corros. Sci., 47 (2005), pp. 2915–2931

96. M. Lebrini, M. Lagrenée, H. Vezin, L. Gengembre, F. Bentiss Electrochemical and quantum chemical studies of new thiadiazole derivatives adsorption on mild steel in normal hydrochloric acid medium Corros. Sci., 47 (2005), pp. 485–505

97. L.-G. Qiu, A.-J. Xie, Y.-H. Shen The adsorption and corrosion inhibition of some cationic gemini surfactants on carbon steel surface in hydrochloric acid Corros. Sci., 47 (2005), pp. 273–278

98. V.R. Saliyan, A.V. Adhikari Quinolin-5-ylmethylene-3-{[8-(trifluoromethyl) quinolin-4-yl]thio}propanohydrazide as an effective inhibitor of mild steel corrosion in HCl solution Corros. Sci., 50 (2008), pp. 55–61

99. A.R. Sathiya Priya, V.S. Muralidharan, A. Subramania Development of novel acidizing inhibitors for carbon steel corrosion in 15% boiling hydrochloric acid Corrosion, 64 (2008), pp. 541–552

100. P.B. Raja, M.G. Sethuraman Natural products as corrosion inhibitor for metals in corrosive media—a review Mater. Lett., 62 (2008), pp. 113–116

101. A. Ostovari, S.M. Hoseinieh, M. Peikari, S.R. Shadizadeh, S.J. Hashemi Corrosion inhibition of mild steel in 1 M HCl solution by henna extract: a comparative study of the inhibition by henna and its constituents (Lawsone, Gallic acid, α-d-Glucose and Tannic acid) Corros. Sci., 51 (2009), pp. 1935–1949

102. A.K. Satapathy, G. Gunasekaran, S.C. Sahoo, K. Amit, P.V. Rodrigues Corrosion inhibition by Justicia gendarussa plant extract in hydrochloric acid solution Corros. Sci., 51 (2009), pp. 2848–2856

103. H. Ashassi-Sorkhabi, M.R. Majidi, K. Seyyedi Investigation of inhibition effect of some amino acids against steel corrosion in HCl solution Appl. Surf. Sci., 225 (2004), pp. 176–185

104. E.R. Fischer, J.E. Parker Technical note: tall oil fatty acid anhydrides as corrosion inhibitor intermediates Corrosion, 53 (1997), pp. 62–64

105. F. Ropital Current and future corrosion challenges for a reliable and sustainable development of the chemical, refinery, and petrochemical industries Mater. Corros., 60 (2009), pp. 495–500

106. M.A. Malik, M.A. Hashim, F. Nabi, S.A. AL-Thabaiti Anti-corrosion ability of surfactants: a review Int. J. Electrochem. Sci., 6 (2011), pp. 1927–1948

107. M.D. Coffey, M.Y. Kelly, W.C. Kennedy, Methods and Composition for Corrosion, in: US Patent 4,493,775, The Dow Chemical Company, 1985.

108. F.B. Growcock, W.W. Frenier Kinetics of steel corrosion in hydrochloric-acid inhibited with trans-cinnamaldehyde J. Electrochem. Soc., 135 (1988), pp. 817–822

109. F.B. Growcock, W.W. Frenier, P.A. Andreozzi Inhibition of steel corrosion in HCl by derivatives of cinnamaldehyde: Part II. Structure–activity correlations Corrosion, 45 (1989), pp. 1007–

1015 N. Fourth North Sea Minister's Conference', Esbjerg, June 8–9, 1995.

110. D.A. Williams, P.K. Holifield, J.R. Looney, L.A. McDougall, Corrosion Inhibitor and Method of Use, in: US Patent 5,130,034, Exxon Chemical Patents Inc., Linden, N.J., 1994.

111. T. Cox, N. Grainger, E.G. Scovell, Oil Production Additive formulations, in: US Patent 2003/0051395 A1, Imperial Chemical Industries Plc, London (GB), United States, 2003.

112. M.M. Brezinski, New environmental options for corrosion inhibitor intensifiers, in: SPE 52707. Presented at the SPE/EPA Exploration and Production Environmental Conference, Austin, TX, February 28–March 3, 1999.

113. M.A. Deyab Effect of cationic surfactant and inorganic anions on the electrochemical behavior of carbon steel in formation water Corros. Sci., 49 (2007), pp. 2315–2328

114. S. Ali, J.S. Reyes, M.M. Samuel, F.M. Auzerais, Self-Diverting Acid Treatment With Formic-Acid-Free Corrosion Inhibitor, in: US patent 2010/0056405 A1, United States, March 4, 2010.

115. D.D.N. Singh, A.K. Dey Synergistic effects of inorganic and organic cations on inhibitive performance of propargyl alcohol on steel dissolution in boiling hydrochloric acid solution Corrosion, 49 (1993), pp. 594–600

116. W.W. Frenier, F.B. Growcock, Mixtures of α,β-Unsaturated Aldehydes and Surface Active Agents Used as Corrosion Inhibitors in Aqueous Fluids, in: US Patent 4,734,259, Dowell Schlumberger Incorporated,Tulsa, Okla, 1988.

117. V.S. Sastri Corrosion Inhibitors: Principles and Applications John Wiley & Sons, Chichester (2001) p. 739

118. T. Kumar, S. Vishwanatham, S.S. Kundu A laboratory study on pteroyl-l-glutamic acid as a scale prevention inhibitor of calcium carbonate in aqueous solution of synthetic produced water J. Petrol. Sci. Eng., 71 (2010), pp. 1–7

119. S. Nešić Key issues related to modelling of internal corrosion of oil and gas pipelines – a review Corros. Sci., 49 (2007), pp. 4308–4338

120. A.F. Beale, C.H. Kucera, Corrosion Inhibitors for Aqueous Acids, in: US Patent 3,231,507, The Dow Chemical Company, Midland, 1966.

121. F.B. Growcock, V.R. Lopp Film formation on steel in cinnamaldehyde-inhibited hydrochloric acid Corrosion, 44 (1988), pp. 248–254

122. F.B. Growcock Inhibition of steel corrosion in HCl by derivatives of cinnamaldehyde: Part I. Corrosion inhibition model Corrosion, 45 (1989), pp. 1003–1007

123. J. Gao, Y. Weng, S. Salitanate, L. Feng, H. Yue Corrosion inhibition of α, β-unsaturated carbonyl compounds on steel in acid medium Petrol. Sci., 6 (2009), pp. 201–207

124. V.S. Sastri Corrosion Inhibitors: Principles and Applications John Wiley & Sons, Chichester (2001) p. 747

125. D.A. Williams, P.K. Holifield, J.R. Looney, L.A. McDougall, Method of Inhibiting Corrosion in Acidizing Wells, in: US Patent 5,089,153, 1992.

126. T. Kumar, S. Vishwanatham Emranuzzaman, study on corrosion control of N80 steel in acid medium using mixed organic inhibitors Indian J. Chem. Technol., 15 (2008), pp. 221–227

127. V.S. Sastri Corrosion Inhibitors: Principles and Applications John Wiley & Sons, Chichester (2001) p. 759

128. S. Ghareba, S. Omanovic The effect of electrolyte flow on the performance of 12-aminododecanoic acid as a carbon steel corrosion inhibitor in CO_2-saturated hydrochloric acid Corros. Sci., 53 (2011), pp. 3805–3812

129. S.C. Baker, D.P. Kelly, J.C. Murrell Microbial-degradation of methanesulfonic-acid – a missing link in the biogeochemical sulfur cycle Nature, 350 (1991), pp. 627–628

130. M.D. Gernon, M. Wu, T. Buszta, P. Janney Environmental benefits of methanesulfonic acid. Comparative properties and advantages Green Chem., 1 (1999), pp. 127–140

131. S. Papavinasam, R.W. Revie, M. Attard, A. Demoz, K. Michaelian Comparison of techniques for monitoring corrosion inhibitors in oil and gas pipelines Corrosion, 59 (2003), pp. 1096–1111

132. S. Papavinasam, R.W. Revie, M. Attard, A. Demoz, K. Michaelian Comparison of laboratory methodologies to evaluate corrosion

inhibitors for oil and gas pipelines Corrosion, 59 (2003), pp. 897–912

133. G 31-72 Standard Practice for Laboratory Immersion Corrosion Testing of Metals, ASTM International, 2004.

134. B.B. Paty, D.D.N. Singh Solvents' role on HCl-induced corrosion of mild steel: its control by propargyl alcohol and metal cations Corrosion, 48 (1992), pp. 442–446 Recommended Practice for Field Testing Water-Based Drilling Fluids, in: Annex E, American Petroleum Institute, ANSI/API RP 13B-1, p. 57–60.

135. G1 – 03 Standard Practice for Preparing, Cleaning, and Evaluating Corrosion Test Specimens, ASTM International, 2003.

136. G 46–94 Standard Guide for Examination and Evaluation of Pitting Corrosion1, ASTM International, 2005.

137. D.L. Massart, B.G.M. Vandeginste, L.M.C. Buydens, S.D. Jong, P.J. Lewi, J. Smeyers-Verbeke Handbook of Chemometrics and Qualimetrics: Part A Elsevir, Amsterdam (1997)

138. E. Poorqasemi, O. Abootalebi, M. Peikari, F. Haqdar Investigating accuracy of the Tafel extrapolation method in HCl solutions Corros. Sci., 51 (2009), pp. 1043–1054

139. E. Gileadi, E. Kirowa-Eisner Some observations concerning the Tafel equation and its relevance to charge transfer in corrosion Corros. Sci., 47 (2005), pp. 3068–3085

140. J.R. Scully Polarization resistance method for determination of instantaneous corrosion rates Corrosion, 56 (2000), pp. 199–218

141. U. Lotz, L. van Bodegom, C. Ouwehand The effect of type of oil or gas condensate on carbonic acid corrosion Corrosion, 47 (1991), pp. 635–645

142. Z.D. Cui, S.L. Wu, C.F. Li, S.L. Zhu, X.J. Yang Corrosion behavior of oil tube steels under conditions of multiphase flow saturated with super-critical carbon dioxide Mater. Lett., 58 (2004), pp. 1035–1040

143. J. Carew, A. Hashem CO_2 corrosion of L-80 steel in simulated oil well conditions Corrosion (2002) Paper No. 02299

144. E. Sadeghi Meresht, T. Shahrabi Farahani, J. Neshati 2-Butyne-1,4-diol as a novel corrosion inhibitor for API X65 steel pipeline in carbonate/bicarbonate solution Corros. Sci., 54 (2012), pp. 36–44

145. E. Naderi, A.H. Jafari, M. Ehteshamzadeh, M.G. Hosseini Effect of carbon steel microstructures and molecular structure of two new Schiff base compounds on inhibition performance in 1 M HCl solution by EIS Mater. Chem. Phys., 115 (2009), pp. 852–858

146. T. Hong, W.P. Jepson Corrosion inhibitor studies in large flow loop at high temperature and high pressure Corros. Sci., 43 (2001), pp. 1839–1849

147. S. Nešić, J. Postlethwaite, S. Olsen An electrochemical model for prediction of corrosion of mild steel in aqueous carbon dioxide solutions Corrosion, 52 (1996), pp. 280–294

148. F.S. de Souza, A. Spinelli Caffeic acid as a green corrosion inhibitor for mild steel Corros. Sci., 51 (2009), pp. 642–649

149. M.G. Hosseini, M.R. Arshadi Study of 2-butyne-1,4-diol as acid corrosion inhibitor for mild steel with electrochemical, infrared and AFM techniques Int. J. Electrochem. Sci., 4 (2009), pp. 1339–1350

150. S. Zhang, Z. Tao, S. Liao, F. Wu Substitutional adsorption isotherms and corrosion inhibitive properties of some oxadiazol-triazole derivative in acidic solution Corros. Sci., 52 (2010), pp. 3126–3132

151. Z. Tao, S. Zhang, W. Li, B. Hou Corrosion inhibition of mild steel in acidic solution by some oxo-triazole derivatives Corros. Sci., 51 (2009), pp. 2588–2595

152. N.O. Obi-Egbedi, I.B. Obot Xanthione: a new and effective corrosion inhibitor for mild steel in sulphuric acid solution Arab. J. Chem., 6 (2013), pp. 211–223

153. M.M. Osman, A.M.A. Omar, A.M. Al-Sabagh Corrosion inhibition of benzyl triethanol ammonium chloride and its ethoxylate on steel in sulphuric acid solution Mater. Chem. Phys., 50 (1997), pp. 271–274

154. L.E. Tsygankova, V.I. Vigdorovich, V.I. Kichigin, E. Kuznetsova Inhibition of carbon steel corrosion in media with H2S studied by impedance spectroscopy method Surf. Interface Anal., 40 (2008), pp. 303–306

155. M.A. Amin, K.F. Khaled, S.A. Fadl-Allah Testing validity of the Tafel extrapolation method for monitoring corrosion of cold

rolled steel in HCl solutions – experimental and theoretical studies Corros. Sci., 52 (2010), pp. 140–151

156. M.A. Amin, K.F. Khaled, Q. Mohsen, H.A. Arida A study of the inhibition of iron corrosion in HCl solutions by some amino acids Corros. Sci., 52 (2010), pp. 1684–1695

Chapter 2

Influence of Temperature on the Atmospheric Corrosion of the Mg–Al Alloy Am50

M. Esmaily[a,] M. Shahabi-Navid[a], J.-E. Svensson[a], M. Halvarsson[b], L. Nyborg[c], Y. Cao[c], and L.-G. Johansson[a]

[a]Department of Chemical and Biological Engineering, Chalmers University of Technology, SE-412 96 Göteborg, Sweden

[b]Department of Applied Physics, Chalmers University of Technology, SE-412 96 Göteborg, Sweden

[c]Department of Materials & Manufacturing Technology, Chalmers University of Technology, 2A, SE-412 96 Göteborg, Sweden

ABSTRACT

The effect of temperature on the NaCl-induced atmospheric corrosion of the Mg–Al alloy AM50 has been investigated in the laboratory. The corroded samples were analyzed gravimetrically and by SEM, EDX, XRD, and FTIR. The atmospheric corrosion of 99.97% Mg was also studied for reference. While the NaCl-induced atmospheric corrosion of AM50 is strongly reduced with decreasing temperature, 99.97% Mg does not exhibit such a trend. The temperature dependence of the atmospheric corrosion of alloy AM50 is attributed to the aluminum content in the alloy. Several crystalline magnesium hydroxy carbonates formed at 4 and 22 °C but were absent at −4 °C.

INTRODUCTION

Magnesium–aluminum (Mg–Al) alloys are widely used as structural materials, e.g., in the aerospace and automotive industries as a result of their light weight, good castability and good mechanical properties at room temperature. However, the use is limited by their relatively poor corrosion resistance [1], [2], [3], [4], [5], [6], [7] and [8]. In general, the corrosion properties of Mg–Al alloys rely on the properties of the partially protective surface film which is made up mainly of $Mg(OH)_2$ but which also contains Al^{3+} [9] and [10]. It is known that the corrosion of Mg and its alloys in aqueous solution is an electrochemical process where magnesium dissolves anodically, the dominant cathodic reaction being hydrogen evolution [11], [12], [13], [14], [15] and [16]. The partially protective surface film is an electronic insulator and the rate of corrosion is strongly accelerated by the presence of cathodically active surfaces such as noble precipitates in the alloy (e.g., MgFe intermetallics) [2], [17], [18], [19], [20] and [21]. Consequently, the concentration of selected impurities (e.g. Fe) is kept below certain limits in commercial alloys in order to avoid forming such precipitates.

Because Mg–Al alloys are often used in atmospheric conditions a better understanding of the atmospheric corrosion properties is of great importance. Accordingly, the atmospheric corrosion properties of Mg and Mg–Al alloys under outdoor conditions have been widely investigated [22], [23], [24], [25] and [26]. The atmospheric corrosion of Mg–Al alloys has also been investigated under controlled conditions in the laboratory, e.g., in order to investigate the mechanism of corrosion and the influence of various parameters on corrosion [27], [28], [29], [30], [31], [32], [33], [34], [35] and [36]. It has been reported that while the rate of corrosion when immersed in aqueous solutions or subjected to salt spray tests is high in comparison to other engineering alloys, the corrosion rates of Mg–Al alloys in the outdoor atmosphere are comparable to mild steel. The causes behind the relatively slow atmospheric corrosion of Mg–Al alloys are still not fully understood [10] and [27]. It is suggested that the inhibitive effect of ambient concentrations of CO_2 on the corrosion of Mg and, Mg–Al alloys reported by [28] and further investigated in a recent study [29] may be part of the explanation.

The corrosion of both Mg and Al in aqueous solution is known to be accelerated by chloride ions [12] and [18]. Hence, the deposition of chlorides in particulate form (e.g. sea salt) is an important factor in the atmospheric corrosion of Mg–Al alloys. In large parts of the world where deicing salt is used, the exterior of cars are heavily exposed to NaCl during winter. Considering that there may be significant NaCl-induced atmospheric corrosion of Mg–Al alloys in automotive applications at temperatures close to or below 0 °C, it is surprising that little scientific work has been devoted to atmospheric corrosion of Mg–Al alloys in that temperature range. Instead, most investigations of atmospheric corrosion in the laboratory are carried out at room temperature or above (e.g. in salt spray testing).

Even when other metallic materials are included, investigations of atmospheric corrosion below ambient temperature are relatively few, especially at sub-zero temperatures [37], [38], [39] and [40]. Working with aluminum at 4–60 °C, Blücher et al. [37] reported on a very strong positive correlation between temperature and the rate of NaCl-induced corrosion in humid air. In contrast, Niklasson et al. [38] reported that the atmospheric corrosion of lead in the presence of gaseous acetic acid exhibited a negative correlation with temperature (4 and 22 °C). Also, Svensson and Johansson [39] reported on a negative correlation with temperature (4 and 22 °C) in the case of SO_2-induced atmospheric corrosion of zinc at high relative humidity. Recently, Chen et al. [40] reported on a positive correlation between corrosion attack and temperature for the initial stages (2 h) of NaCl-induced atmospheric corrosion of the Mg–Al alloy AZ91 between 2°C and 8°C. The work was carried out in pure water vapor at about 0.01 atm and only investigated the initial attack (2h).

The present paper is the first attempt to investigate the temperature dependence of the atmospheric corrosion of an Mg–Al alloy down to sub-zero temperatures. Thus, the NaCl-induced atmospheric corrosion of the Mg–Al alloy AM50 is investigated at 22, 4 and −4 °C. In order to study the effect of temperature rather than the combined effects of temperature and relative humidity, all exposures are carried out at a constant relative humidity of 95%. In addition, we report on the influence of ambient levels of CO_2 on corrosion behavior. 99.97% Mg is used as a reference material and comparisons are made to samples exposed in the absence of NaCl.

· EXPERIMENTAL

Sample Preparation

Chemical compositions of the tested materials, namely pure Mg (99.97%) and alloy AM50, are listed in Table 1. Unalloyed Mg (99.97% Mg) was used as reference material. As-received material was machined to obtain $14 \times 14 \times 3$ mm^3 test coupons with an exposed surface area of 5.56 cm^2 from high purity Mg and High Pressure Die Cast (HPDC) AM50 alloy. Prior to the exposures, samples were ground, lubricated with de-ionized water, through successive grades of silicon carbide abrasive papers (SiC grit papers) from P1000 to P4000 mesh. Polishing was then performed using cloth discs and diamond paste in the range of 1–3 µm, followed by a fine polishing step using oxidized porous silicon (OPS) for 120 s on a Buehler Microcloth. The specimens were cleaned using distilled water and degreased with acetone, washed again with distilled water, and dried with a blower (cool air) and stored in a desiccator over a desiccant for 24 h (h) before exposure. A solution of 20 ml distilled water, 80 ml ethanol and 1 g NaCl was sprayed on the test coupons. The samples were contaminated with two different amounts of salt, 14 and 70 µg/cm^2. Care was taken to avoid an uneven distribution of NaCl on the specimen surface by the spraying of the NaCl solution.

Table 1: Composition of the test materials (% by weight)

	Al	Zn	Mn	Si	Fe	Cu	Ni	Ca
Mg	0.0030	0.0050	0.0023	0.0030	0.0018	0.0003	0.0002	0.0010
AM50	5.0	0.01	0.25	0.01	0.0016	0.0010	0.0007	n.a.

In order to achieve an even distribution of salt, the spraying procedure was optimized with respect e.g. to the distance between the spray gun and by dividing the spraying into several steps with intermittent drying. Also, reference samples, i.e. samples without salt, were exposed at the same time with the contaminated samples in each corrosion condition. Duplicate and triplicate samples were exposed for each condition and some of the experiments were repeated several times to check for consistency. In total about 450 samples were exposed.

Experimental Set-Up

The experimental set-up for the exposures with 400 ppm CO_2 has been described in detail previously [27] and [28]. The exposure apparatus is entirely made in glass and Teflon. The samples are suspended by a nylon string in individual exposure chambers. The gas flow is 1000 ml/min corresponding to an average flow velocity of 1 mm/s. Relative humidity (RH) is regulated by mixing measured amounts of dry air and air saturated with water vapor at the exposure temperature. RH was 95 ± 0.3%. CO_2 is added from a cylinder to give a constant concentration of 400 ± 20 ppm. The amount of CO_2 added was measured before and after the experiment. To perform corrosion experiment at sub-zero temperature, a new corrosion system was designed and developed, as illustrated in Fig. 1. Ethylene glycol was used as antifreeze. In addition, two internal humidifiers were placed inside the liquid of the pool to decrease the dew point of the humidified air to −4 °C.

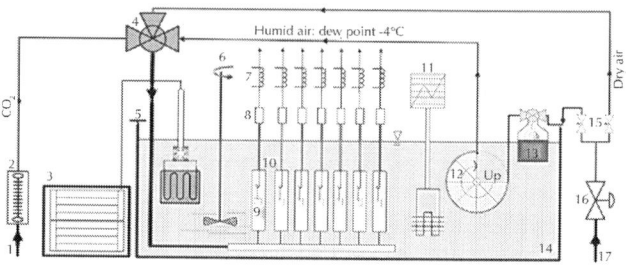

Figure 1: Schematic diagram of the exposure chamber for the sub-zero atmospheric corrosion experiment: (1) CO_2 source, (2) flow meter, (3) dip cooler, (4) mixing chamber, (5) insulation, (6) stirrer, (7) solenoid valves, (8) wash bottles, (9) corrosion samples suspended by nylon string, (10) corrosion chambers, (11) temperature regulator, (12 and 13) humidifiers producing 95% RH air at the exposure temperature (-4 °C), (14) water +44% ethylene glycol at constant temperature, (15) needle valves, (16) manometer valve (17) dry purified air with a pressure of 6 bars.

The corrosion exposures in the absence of CO_2 were carried out in hermetically closed desiccators (as corrosion reactors) with a volume of $3 dm^3$. Exposure temperatures were 22.00 ± 0.03 °C 4.00 ± 0.03 °C and -4.00 ± 0.03 °C. The 95% humid air was achieved by equilibrating with 500 ml potassium hydroxide solution (KOH (aq)) in a container placed inside the desiccator. The water vapor partial pressure over KOH (aq) was calculated by the equation reported in [41]. In addition to the RH to the desired value, i.e. 95%, the KOH (aq) solution also functions as a CO_2 absorbent.

Quantitative Measurements

Gravimetric Measurement

To monitor the atmospheric corrosion process the test coupons were removed from reactors during the exposures for weighing with regular intervals except for the CO_2-free exposures, i.e. desiccator exposures. Therefore, the water contained in the NaCl electrolyte remains during weighing and the corresponding mass gains are accordingly called wet mass gains. At the conclusion of an exposure, after measuring the wet mass, the samples were stored for 24 h at room temperature over a desiccant so that the loosely bound water was removed. The specimens were weighed again and the corresponding mass gains are termed dry mass gains.

Leaching and Pickling

The quantity of corrosion product was determined via leaching and pickling processes of the corroded specimens using agitating ultrasonic bath at room temperature. The water soluble corrosion products and unreacted NaCl were first removed by leaching process. To do this, the samples were immersed in Mili-Q water (ultrapure water) in two steps; 30 s and 60 s at 25 °C. Subsequently, the corroded samples were pickled several times in a chromate solution of 20% chromium trioxide (CrO_3) for 15 s followed by several periods of 30 s. The heavily corroded samples were pickled up to 25 times. The samples were cleaned by pure water and acetone and finally dried by a stream of cold air after each step. The weight of the samples was registered after each leaching and pickling process to monitor the removal process of the corrosion products. The metal loss was determined by weighing the samples after leaching and pickling. It is noticed that self-corrosion during pickling was negligible. Quantitative data on corrosion product stoichiometry were obtained by calculating the corrosion product ratio according to the following equations: equation (1).

$$_M loss = (_{MO}) - (_{MP})$$ (1)

$$_M Corr = (_M Dry) - (_{MP})$$ (2)

$$X = (_M Corr) - (_M loss)$$ (3)

where M_{loss} is metal loss, M_O is mass before exposure and salt, M_P is mass after final stage of pickling, M_{Corr} is mass of corrosion product, M_{Dry} is dry mass and X specifies corrosion product ratio. It should be mentioned that the gravimetric values were averaged from the data (3–8 samples for each test condition) with error bars based on one standard deviation. Some of the exposures were repeated and in total 700 samples were prepared, sprayed, exposed and analyzed to different levels. All gravimetric data were obtained using a Sartorius Microbalance with a with 0.0001 mg resolution.

Analytical Techniques

X-Ray Diffraction (XRD)

Crystalline corrosion products formed under various exposure conditions were analyzed by X-ray diffraction (XRD) using a Bruker AXS D8 powder diffractometer. The system was equipped with grazing incidence beam attachment and a Göbel mirror. Cr K radiation (= 2.29 Å) was used and the angle of incidence was 5°.

Fourier Transform Infrared Spectroscopy (FTIR)

In addition to XRD, the corrosion products were also analyzed by Fourier transform infrared (FTIR) spectroscopy. The investigations were conducted from 4000 to 400 cm^{-1} to identify the functional groups in the corrosion products. The FTIR spectrometer used in this study was a Nicolet Magna-IR 560 equipped with a PTGS detector with an insert cell for diffuse reflectance spectroscopy.

Analytical Scanning Electron Microscope (SEM/EDX)

The morphology of the corrosion products was examined by an FEI Quanta 200 environmental scanning electron microscopy (ESEM) with a Schottky field emission gun (FEG) both in the plane view and ion milled cross section investigations (see below). The instrument was equipped with Oxford Inca energy dispersive X-ray detector (EDX) system. Chemical composition analysis was performed with an Oxford Inca energy dispersive X-ray system (EDX). Imaging was performed using a range of acceleration voltages, 5–20 kV. SEM/EDX was used for local chemical analysis as well as elemental mapping of the corroded metal surfaces and the ion milled cross sections.

Focused Ion Beam Milling (Fib) and Broad Ion Beam Milling (Bib)

The distribution of elements after exposures was studied on the cross sections of samples prepared by broad ion beam milling (BIB) and focused ion beam milling (FIB) methods using a Leica EM TIC 3X ion beam slope cutter and an FEI Versa 3D system, respectively. The BIB cross sectioning is a relatively advanced technique that can be used to produce accurate cross sections with limited artifacts and distortion through the corrosion product and metal substrate in order to study the fragile corrosion layers.

Considering the larger surface area (around 1.5 mm in width and several hundred microns deep) of the cross sections produced by BIB in comparison with those of the FIB, the method was well-suited to study the corrosion pits on the heavily corroded samples. In the case of the BIB the sample holder was continuously cooled using liquid nitrogen in order to avoid any compositional change in the corrosion products. An argon ion beam with an acceleration voltage of 3 kV was chosen for the milling process. The cross sections were mainly made by BIB, but in the case of samples with very low amount of corrosion products, when precise positioning of the cross section was required, the FIB technique was used. The instrument is a dual beam system, equipped with both an electron and ion column. The electron column is equipped with a field emission gun (FEG) and the ion column with a liquid gallium source. Before milling, a 2 μm nickel layer was deposited by physical vapor deposition (PVD) onto the corroded samples, in order to protect the surface oxide from ion-induced damage.

RESULTS

Gravimetric Measurements

The dry mass gains as a function of exposure time for both materials are presented in Fig. 2 and Fig. 3 and Table 2 and Table 3. The mass gain and metal loss data for the reference samples (exposed in the absence of salt) was in most cases negligible (<1 µm/year) and are not shown. All the gravimetric data, including the mass gain and metal loss values, are provided to enable quantitative comparisons of the influence of the different environmental factors on the atmospheric corrosion of the alloy AM50. Mass gain is a convenient way to measure average corrosion attack if the corrosion product composition is known. Also, mass gains can be used to compare corrosion rates if the corrosion products have similar composition. The tables also show the average metal loss of the corroded samples. The ratio between the mass of the corrosion products and the metal loss for individual samples is also given. This value reflects the average corrosion product composition and can be compared to the corresponding ratios of selected corrosion products listed in Table 4. When the mass gain was <0.05 mg cm^{-2} metal loss could not be determined with sufficient precision and metal loss was calculated from the mass gain values, using the appropriate ratios of corrosion products identified by XRD (Table 4).

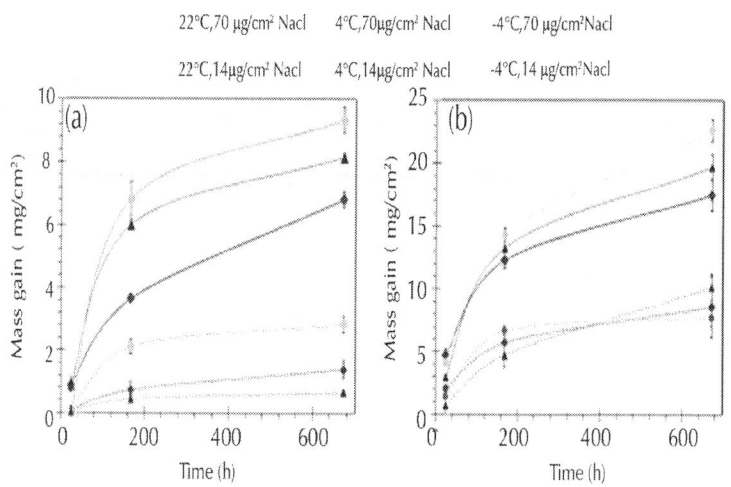

Figure 2: The effect of temperature and exposure time on mass gain of 99.97% Mg: (a) 400 ppm CO_2. (b) CO_2-free. The standard errors for the data in Fig. 2a and b is less than 5 and 10%, respectively.

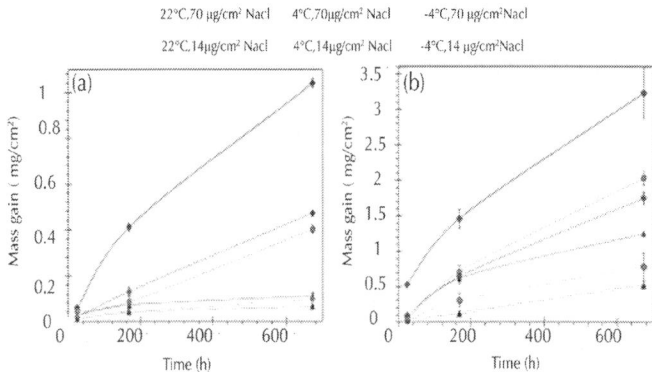

Figure 3: The effect of temperature and exposure time on the mass gain of AM50: (a) 400 ppm CO_2. (b) CO_2-free. The standard errors for the data in Fig. 3a and b is less than 3 and 10%, respectively.

Table 2: Metal losses, corrosion rates and experimental corrosion product ratios for 99.97% Mg

CO_2	Temperature	NaCl (μg/cm²)	Exposure time (h)	Metal loss (mg/cm²)	Ratio (X) [a]	Corrosion rate (μm/year)
400 ppm	22	14	24	0.08 ± 0.02	2.1	
400 ppm	22	14	168	0.32 ± 0.04	3.3	
400 ppm	22	14	672	0.50 ± 0.02	3.8	52 ± 1
400 ppm	4	14	24	0.04 ± 0.01	4.0	
400 ppm	4	14	168	1.11 ± 0.21	2.9	
400 ppm	4	14	672	0.58 ± 0.12	2.4	64 ± 4.2
400 ppm	-4	14	24	0.12 ± 0.06	1.3	
400 ppm	-4	14	168	0.29 ± 0.09	2.5	

400 ppm	−4	14	672	0.37 ± 0.08	2.8	27 ± 3
400 ppm	22	70	24	0.41 ± 0.02	2.8	
400 ppm	22	70	168	1.99 ± 0.14	2.8	
400 ppm	22	70	672	3.37 ± 0.32	3.0	239 ± 21
400 ppm	4	70	24	0.01 ± 0.005	2.0	
400 ppm	4	70	168	3.83 ± 0.34	2.8	
400 ppm	4	70	672	5.18 ± 0.83	2.8	355 ± 46
400 ppm	−4	70	24	0.51 ± 0.05	2.7	
400 ppm	−4	70	168	3.15 ± 0.31	2.9	
400 ppm	−4	70	672	2.68 ± 0.43	2.9	199 ± 28
0	22	14	24	1.40 ± 0.23	2.5	
0	22	14	168	3.40 ± 0.29	2.7	
0	22	14	672	5.70 ± 0.48	2.5	546 ± 27
0	4	14	24	0.89 ± 0.21	2.5	
0	4	14	168	2.27 ± 0.33	2.5	
0	4	14	672	4.30 ± 0.54	2.7	294 ± 32
0	−4	14	24	0.55 ± 0.08	2.3	
0	−4	14	168	3.33 ± 0.19	2.4	
0	−4	14	672	7.53 ± 0.95	2.3	601 ± 49
0	22	70	24	3.45 ± 0.84	2.3	
0	22	70	168	7.03 ± 1.22	2.7	
0	22	70	672	15.50 ± 2.34	3.6	1161 ± 64

0	4	70	24	2.61 ± 0.24	2.6	
0	4	70	168	6.91 ± 1.12	3.1	
0	4	70	672	12.41 ± 1.45	2.8	773 ± 58
0	−4	70	24	2.14 ± 0.44	2.4	
0	−4	70	168	7.26 ± 1.2	2.8	
0	−4	70	672	12.74 ± 2.23	2.5	993 ± 49

[a]Ratio of the dominant corrosion product detected by XRD

Table 3: Metal losses, corrosion rates and experimental corrosion product ratios for the alloy AM50

CO_2	Temperature	NaCl (µg/cm²)	Exposure time (h)	Metal loss (mg/cm²)	Ratio (X)	Corrosion rate (µm/year)
400 ppm	22	14	24	0.01[b]	4.0	
400 ppm	22	14	168	0.04 ± 0.001	4.2	
400 ppm	22	14	672	0.15 ± 0.02	4.1	11 ± 1
400 ppm	4	14	24	0.01[b]	4.0	
400 ppm	4	14	168	0.01[b]	4.0	
400 ppm	4	14	672	0.03 ± 0.01	4.0	2 ± 1
400 ppm	−4	14	24	<0.01[b]	[a]4.0	
400 ppm	−4	14	168	0.01[b]	4.0	
400 ppm	−4	14	672	0.04 ± 0.001	2.5	3 ± 1
400 ppm	22	70	24	0.03 ± 0.001	4.0	
400 ppm	22	70	168	0.13 ± 0.003	4.2	

400 ppm	22	70	672	0.34 ± 0.002	4.1	25 ± 1
400 ppm	4	70	24	0.01[b]	4.0	
400 ppm	4	70	168	0.02[b]	5	
400 ppm	4	70	672	0.13 ± 0.006	4.1	10 ± 1
400 ppm	−4	70	24	0.01[b]	2.5	
400 ppm	−4	70	168	0.03 ± 0.001	3.3	
400 ppm	−4	70	672	0.05 ± 0.001	3.2	4 ± 1
0	22	14	24	0.12 ± 0.003	1.7	
0	22	14	168	0.15 ± 0.002	2.7	
0	22	14	672	1.08 ± 0.14	2.6	78 ± 4.5
0	4	14	24	0.04 ± 0.001	[a]2.4	
0	4	14	168	0.20 ± 0.05	2.5	
0	4	14	672	0.46 ± 0.12	2.7	45 ± 4.2
0	−4	14	24	<0.01[b]	[a]2.4	
0	−4	14	168	0.10 ± 0.002	2.3	
0	−4	14	672	0.34 ± 0.09	2.6	26 ± 1
0	22	70	24	0.36 ± 0.07	2.5	
0	22	70	168	0.90 ± 0.12	2.6	
0	22	70	672	2.10 ± 0.19	2.5	124 ± 5.2
0	4	70	24	0.03 ± 0.001	3.0	
0	4	70	168	0.43 ± 0.04	2.6	
0	4	70	672	1.23 ± 0.13	2.9	89 ± 4.4
0	−4	70	24	0.01[b]	2.3	

| 0 | -4 | 70 | 168 | 0.20 ± 0.002 | 2.6 | |
| 0 | -4 | 70 | 672 | 0.72 ± 0.11 | 2.7 | 53 ± 4.2 |

[a]Ratio of the dominant corrosion product detected by XRD.

[b]The scatter for these cases were negligible ($<\pm0.001$ mg/cm^2)

Table 4: Crystalline phases identified by XRD. Two corrosion products are included (in italics) that were not positively identified but are considered possible candidates

Name	Formula	Abbreviation	Corrosion product ratio (total mass/mass of metal ions)
Magnesium	-Mg	Mg	–
-phase	$Mg_{17}Al_{12}$		–
Brucite	$Mg(OH)_2$	B	2.40
Hydromagnesite (1)	$Mg_5(CO_3)_4(OH)_2 \times 4H_2O$	H4	3.85
	$Mg_5(CO_3)_4(OH)_2 \times 5H_2O$	H5	4.00
Giorgiosite	$Mg_5(CO_3)_4(OH)_2 \times 8H_2O$	H8	4.44
Meixnerite	$Mg_6Al_2(OH)_{18} \times 4.5H_2O$	Meix	4.02
Hydrotalcite	$Mg_6Al_2(OH)_{16}CO_3 \times 4H_2O$		*4.20*
	$Mg_2Cl_2CO_3 \times 7H_2O$		*4.82*

The Effect of NaCl and Exposure Time

In the absence of NaCl, corrosion rates were very small, the mass gains registered after 4 weeks corresponding to an average corrosion rate of AM50 of <1 µm/year. The mass gains of unalloyed Mg are somewhat higher and the largest mass gain in the absence of salt was registered for 99.97% Mg after 672 h at 22 °C, in the absence of CO_2. As expected, the mass gains were much greater in the presence of salt, reflecting the corrosivity of NaCl towards Mg and Mg alloys under humid conditions. Fig. 2 and Fig. 3 show that most mass gain curves are convex, corresponding to a gradual slowing of the corrosion process. However, in the case of AM50 in the absence of CO_2 at −4 °C, the mass gain curve was linear.

Comparing 99.97% Mg and AM50

The mass gains of 99.97% Mg were considerably larger than those of the alloy, independent of the amount of salt, temperature and exposure time (Fig. 2 and Fig. 3). Also, the mass gain values of alloy AM50, relative to 99.97% Mg, increased with decreasing the temperature. For example, in the absence of CO_2 with 70 μg cm^{-2}salt, the mass gains of 99.97% Mg after 672 h were 6 times larger than those of AM50 at 22 °C while the corresponding factor was 16 at −4 °C. In the presence of CO_2, the corresponding factors were 7 at 22 °C and 50 at −4 °C. The metal loss data showed the same trends (see Table 2 and Table 3).

The Effect of Co_2

According to a previous study by this group [29], CO_2 inhibits the NaCl-induced atmospheric corrosion of 99.97% Mg and AM50 at 22 °C. The present results illustrate the strong inhibitive effect of CO_2 towards both pure Mg and the alloy at all three temperatures; see Fig. 4a and b. At 4 and −4 °C, the metal loss values of 99.97% Mg in the absence of CO_2 were 2–20 times greater than in the presence of 400 ppm CO_2 (Table 2). The corrosion inhibitive effect of CO_2 was even stronger for the alloy. The metal loss values of AM50 in the absence of CO_2 were greater by a factor of 9–20 (4 °C) and 6–14 (−4 °C) than in the presence of 400 ppm of CO_2 (Table 3). It should be noted that metal loss data should be used rather than mass gain data when investigating the corrosion inhibitive effect of CO_2 as the corrosion product composition is strongly influenced by CO_2.

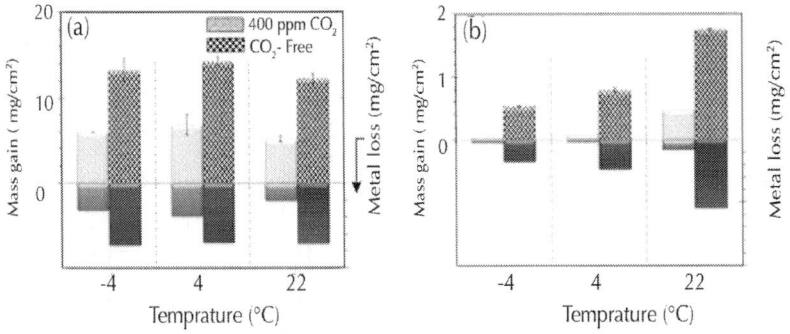

Figure 4: The effect of temperature on the mass gain and metal loss (70 µg cm^{-2} NaCl, 168 h) (a) 99.97% Mg, (b) AM50.

The Effect of Temperature

In order to illustrate the influence of temperature on atmospheric corrosion behavior of the investigated alloys, the dry mass gain and metal loss values after 168 h as function of exposure temperatures are plotted in Fig. 4. The corresponding data for the cases of 24 and 672 h are listed in Table 2 and Table 3. The mass gain and metal loss results after 168 h show that the effect of temperature on the rate of NaCl-induced corrosion of 99.97% Mg was different from the case of alloy AM50, see Fig. 4. For the alloy, increasing the temperature from −4 °C to 22 °C significantly increased the average rate of corrosion, irrespective of the amount of added NaCl and both in the absence and in the presence of CO_2, as shown in Fig. 2. The effect of temperature on alloy AM50 was especially marked in the presence of CO_2. Hence, between −4 and 22 °C, the rate of NaCl-induced corrosion of AM50 increased by a factor of about 6 and 4.5 in the presence and absence of CO_2, respectively (Table 3 and Fig. 4b). In the case of 99.97% Mg, there was a negative correlation between corrosion rate and temperature in the presence of CO_2, corrosion was significantly slower at 22 °C compared to 4 and −4 °C (Fig. 4a). In contrast, the effect of temperature was not significant in the absence of CO_2.

Corrosion Product Composition

The phases identified by XRD on samples and also the ratios between formula mass and the mass of the metal ions for the corrosion products are presented in Table 4. The crystalline phases identified after exposures at three temperatures are shown in Table 5. The crystalline corrosion products at 4 and 22 °C were the same. In the absence of both NaCl and CO_2, traces of brucite were identified on 99.97% Mg but not on AM50. In the absence of salt and in the presence of CO_2, no crystalline products were found. NaCl-induced corrosion in the absence of CO_2 resulted in the formation of brucite on both materials. In the case of the alloy, the magnesium aluminum hydroxide meixnerite ($Mg_6Al_2(OH)_{18}$ $\times 4.5H_2O$) was also identified. In the presence of CO_2 and NaCl, both materials produced three magnesium hydroxy carbonates; H4, H5 and H8 (Table 4). In some cases, the diffractograms contained additional weak peaks that could not be attributed to a known compound.

Table 5: Phases identified by XRD after exposure at the three temperature. Abbreviations are explained in Table 4. The presence of an unidentified crystalline corrosion product is indicated by the letter U and U1

Material	CO_2 (ppm)	NaCl (µg/cm²)	Exposure time (h)	Crystalline phases detected		
				From base alloy	4 and 22 °C	−4 °C
AM50	400	0	All	Mg(s), (w)		U1 (after 672 h)
AM50	400 ppm	70	24	Mg(s), (w), NaCl(s)		
AM50	400 ppm	70	672	Mg(s), NaCl	H8, H5, H4, U	U1
AM50	0	0	All	Mg(s), (w)		U1 (after 672 h)
AM50	0	70	24	Mg(s), NaCl	B, Meix	B
AM50	0	70	672	Mg(s)	B, Meix	B, Meix, U1
Mg	400 ppm	0	All	Mg(s)		
Mg	400 ppm	70	24	Mg(s), NaCl	B	U

Mg	400 ppm	70	672	Mg(s)	B, H5, H4, H8	B
Mg	0	0	24	Mg(s)		B, U1
Mg	0	0	672	Mg(s)	B	B, U1
Mg	0	70	24	Mg(w)	B	U1
Mg	0	70	672	Mg(w)	B	B, U1

The corrosion products identified after exposure at −4 °C were different from those formed at higher temperatures. Therefore, the only crystalline compounds identified at −4 °C were brucite (on 99.97%Mg and AM50) and meixnerite (on AM50) after exposure with NaCl in the absence of CO_2 (Table 5). Interestingly, the crystalline hydroxy carbonates formed under these conditions at 4 and 22 °C were absent at −4 °C. Instead an unknown phase appeared (U1 in Table 5) in most cases, with reflections at $d = 4.58$ Å and $d = 2.95$ Å. The unknown phase is illustrated by Fig. 5a and b, showing diffractograms acquired from 99.97% Mg and AM50 exposed in the absence of CO_2 at −4 °C. FTIR spectra acquired from samples exposed after 672 h exposure at −4 °C and 4 °C are depicted in Fig. 5d. The absorption bands at 1435, 1488 cm^{-1} on the sample exposed at 4 °C correspond to carbonate, in accordance with the identification of magnesium hydroxy carbonate by XRD analysis (Table 5).

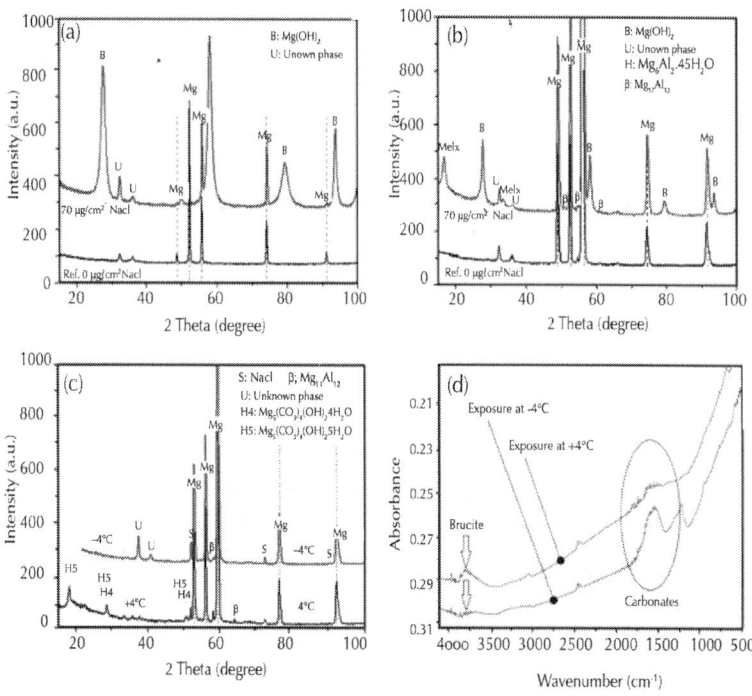

Figure 5: XRD patterns of (a) 99.97% Mg and (b) AM50 exposed for 672 h at −4 °C in the absence of CO_2 and NaCl (lower (black) pattern) and in the presence of 70 μg/cm² NaCl (upper (red) patterns). (c) XRD patterns of AM50 exposed for 672 h at −4 °C and 4 °C in the presence of CO_2 with 70 μg/cm² NaCl. (d) two representative FTIR spectra of AM50 after 672 h at −4 and 4 °C in the presence of CO_2 and with 70 μg/cm² NaCl. (For interpretation of the references to colour in this figure legend, the reader is referred to the web version of this article.)

Microscopy of Corrosion of Alloy AM50

The microstructure of alloy AM50 investigated is described elsewhere [18] and [42]. The main microstructural constituents are shown in the plan-view SEM image and cross section image in Fig. 10a and b. The alloy is composed of -Mg grains and a partially divorced   + eutectic. In addition,   phase (Al_8Mn_5(Fe)) particles are present, usually in the   +   eutectic region. The plan-view SEM images in Fig. 6, Fig. 7, Fig. 8 and Fig. 9show the morphology of the corroded metal surface after exposure. It may be noted that in the present study all corrosion products, including the water soluble ones, remain on the metal surface during the exposures, which is not the case in e.g., immersion testing [15] and [16].

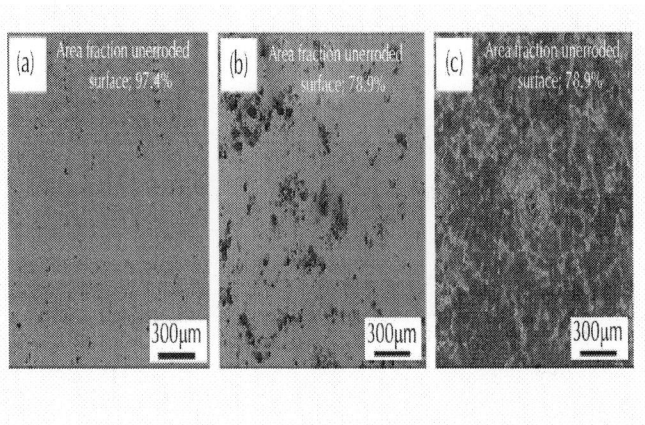

Figure 6: The effect of NaCl on corrosion morphology on alloy AM50, 4 °C, 672 h, 400 ppm CO_2 (a) no NaCl, (b) 14 µg/cm² NaCl, (c) 70 µg/cm² NaCl.

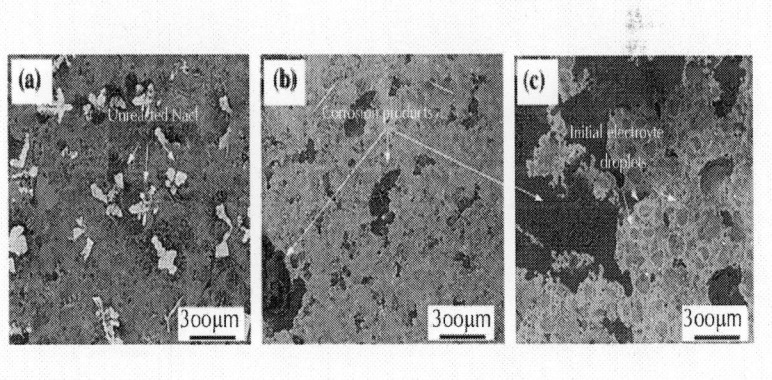

Figure 7: The effect of time on corrosion morphology on alloy AM50, 4 °C, 70 µg/cm² NaCl, CO_2-free (a) 24 h, (b) 168 h and (c) 672 h.

Figure 8: SEM secondary electron images showing the effect of temperature on the corrosion morphology of alloy AM50, 70 μg/cm² NaCl, 672 h (a) 22 °C, 400 ppm CO_2, (b) −4 °C 400 ppm CO_2, (c) 22 °C, CO_2-free and (d) −4 °C, CO_2-free.

Figure 9: SEM secondary electron images showing the corrosion product morphology after 672 h exposure with 70 μg/cm² NaCl in the presence of CO_2; (a) at 22 °C, (b) at −4 °C.

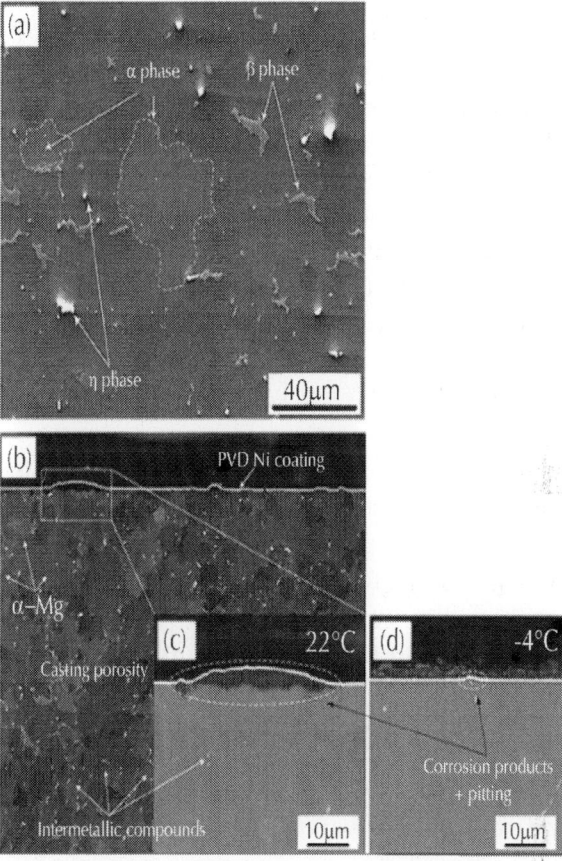

Figure 10: SEM images of (a) metal surface showing the as-cast microstructure of the alloy AM50. (b–d) SEM secondary and backscattered electron images of BIB-prepared cross-sections and plane view image of alloy AM50 exposed in the presence of CO_2 and 14 µg/cm² NaCl after 168 h at 22 and −4 °C. Nickel was deposited on the top surface to protect the corrosion product. The difference in brightness between different -Mg grains is the result of channeling contrast.

The strongly corrosive nature of NaCl is illustrated in Fig. 6. As seen, there is little evidence of corrosion in the absence of salt and the extent of corrosion increases with the amount of salt added. This is in accord with the gravimetric results (Table 2 and Table 3). The sample that had corroded in the presence of 70 µg cm⁻² of salt in the presence of CO_2 exhibits unreacted NaCl after 672 h exposure, which could also be identified by XRD (Table 5). Fig. 7 illustrates the effect of exposure time on corrosion morphology for NaCl contaminated AM50 exposed in the absence of CO_2 at 4 °C. As expected, there was a considerable increase in the extent of corrosion from 24 to 672 h (Fig. 7a and b). After 24 h, corrosion product agglomerations had formed on the surface (grey and dark grey) and large amounts of unreacted NaCl were present. The NaCl crystallites correspond to a NaCl (aq) solution that was present at the end of the exposure and that has crystallized upon drying. After 672 h, NaCl was no longer detected on the surface that was also in accordance with the XRD analysis (Table 5). The roughly circular surface features (as shown in Fig. 7c) of about 20–100 μm diameter correspond to electrolyte droplets that formed at the start of the exposure.

The effect of exposure temperature on the morphology of the corroded surface in the presence and absence of CO_2 at −4 and 22 °C is illustrated in Fig. 8. The corresponding exposures at 4 °C are shown in Fig. 6c and Fig. 7c. In accordance with the gravimetric results in Table 2 and Table 3, there was a strong positive correlation between the extent of corrosion and temperature, both in the absence and in the presence of CO_2. The corrosion inhibitive effect of CO_2 was also evident from the images (compare Fig. 8a, b with c, d (−4 and 22 °C) and Fig. 6 and Fig. 7 (4 °C). Thus, the samples least affected by corrosion are the ones exposed at −4 °C in the presence of CO_2 (Fig. 8b), the image revealing circular corrosion product accumulations on an otherwise smooth surface. The corresponding exposure at 22 °C resulted in a much more strongly corroded surface with corrosion product crusts covering most of the surface. The circular features correspond to NaCl (aq) droplets that formed when the samples were introduced into the humid exposure environment. It may be noted that the morphology is rather different from that described for corrosion products on Mg Al alloys exposed in the outdoor atmosphere [23] and [24]. This is because the latter have been subjected to repeated wet-dry cycling, resulting in the formation of cracks and crack rings around the corrosion crusts.

The strong effect of temperature in the presence of CO_2 is in agreement with the gravimetric results. Exposures in the absence of CO_2 resulted in a very different corrosion morphology, see Fig. 8c and d. Thus, extensive worm-like corrosion products appeared at all three temperatures, the area between the corrosion product accumulations was being relatively unaffected by corrosion. As expected, the most severe corrosion was observed at 22 °C in the absence of CO_2 while the exposure at −4 °C in the presence of 400 ppm CO_2 represented the mildest corrosive conditions. CO_2 had a similar effect on corrosion morphology at all three temperatures, producing a much more uniform corrosion attack compared to the samples exposed in the absence of CO_2.

The corrosion morphology observed in the presence of CO_2 at 4 °C and 22 °C is dominated by plate-like clusters, the plates were being oriented perpendicular to the sample surface (Fig. 9). These corrosion product aggregates were highly porous. Similar corrosion product morphologies have also been reported previously for Mg alloys [35], [43] and [44] and are suggested to correspond to the magnesium hydroxy carbonates.

The samples exposed at −4 °C exhibited a completely different morphology; the corrosion products were being dominated by egg-shell like features, see Fig. 9. It may be noted that the absence of the needle-like corrosion products at −4 °C was accompanied by a lack of evidence for crystalline magnesium hydroxy carbonates by XRD. Compared to the NaCl-induced corrosion of alloy AM50 at 22 °C and 400 ppm CO_2, the corresponding exposure at −4 °C resulted in very few corrosion crusts owing to the slow corrosion rate (see SEM image of BIB-prepared cross section in Fig. 10).

Because of the severe corrosion attack at 22 °C it was possible to find corrosion pits in the BIB-prepared cross sections (Fig. 10d). However, this was not possible at −4 °C because of the mild corrosion attack. To locally investigate the corrosion attack at −4 °C, FIB-prepared cross-sections were used instead. Fig. 11 shows the SEM backscattered electron image and EDX maps of the cross section for Al, C, O, Cl and Mn acquired after 168 h exposure in the presence of CO_2. The image shows a corrosion product of varying thickness and pits in the alloy. The elemental mapping shows that Cl is concentrated in the bottom of the pits. Also, the pits are situated in the vicinity of areas where phase is present (see Al map in Fig. 11). An -phase particle can be seen in the backscattered image and in the Mn map. Al is scarce in the corrosion product which is dominated by Mg. The carbon map shows that the chlorine-rich areas are also rich in carbonate. The sodium signal was not significant because of low concentration and to overlap with magnesium in the EDX spectrum.

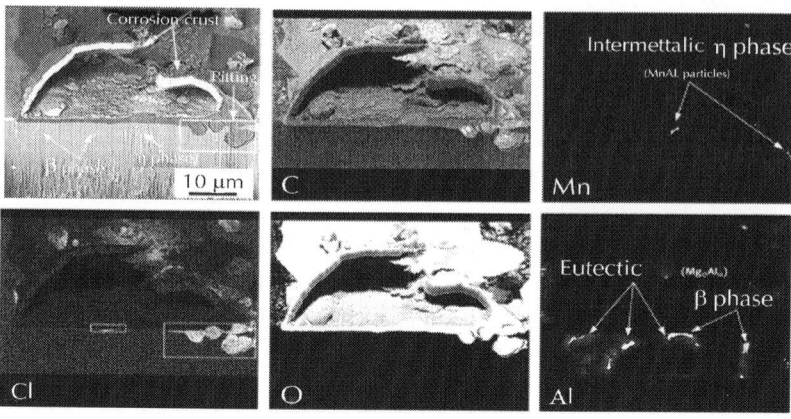

Figure 11: SEM backscattered electron image and EDX maps of FIB-prepared cross-sections of alloy AM50 after 168 h exposure in the presence of 400 ppm CO_2 and 14 $\mu g/cm^2$ NaCl at −4 °C. The rectangles are showing the pitting corrosion and accumulations of Cl⁻ ions.

As noted above, the corrosion of alloy AM50 is more severe and localized in the absence of CO_2 at all temperatures. Fig. 12 shows the top-view of the AM50 alloy after 168 h exposure at −4 °C in the absence of CO_2. The image may be compared to the Fig. 8d, where corrosion is more pronounced both because of longer exposure time (672 h) and higher salt concentration (70 $\mu g/cm^2$). Fig. 12 shows a secondary electron image and a close-up of a relatively uncorroded part of the surface, at a distance from the localized attack. The quantitative elemental analysis of the partly detached corrosion product layer indicates that it is enriched in sodium and poor in chlorine. The elemental analysis indicates that relatively high amounts of carbonate are present. It is suggested that the carbonate has formed during sample handling and that the sodium-rich areas originally consisted of NaOH. The relatively low concentration of Al indicates that meixnerite $(Mg_6Al_2(OH)_{18} \times 4.5H_2O)$ is absent in this area.

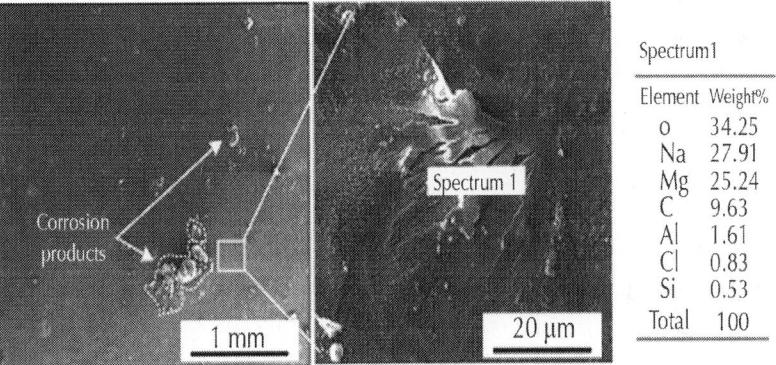

Spectrum1	
Element	Weight%
O	34.25
Na	27.91
Mg	25.24
C	9.63
Al	1.61
Cl	0.83
Si	0.53
Total	100

Figure 12: The EDX analysis shows that sodium is enriched on the surface of the sample and the chlorine level is below the detection limit. Alloy AM50 exposed in the absence of CO_2 and 14 µg/cm² NaCl after 168 h at −4 °C.

DISCUSSION

The gravimetric results (Table 2 and Table 3, Fig. 3 and Fig. 4) and the corrosion morphology (Fig. 6, Fig. 7,Fig. 8 and Fig. 9) illustrate the well-known corrosiveness of NaCl towards Mg and Mg alloys [10] and [11], the rate of atmospheric corrosion being very low (<1 µm/ year) in the absence of salt. The NaCl added before exposure forms an aqueous solution in the experimental conditions (95% RH) at all three temperatures investigated. Accordingly, droplets consisting of NaCl (aq) solution were observed on the surface in the beginning of the exposures. In many cases, circular features corresponding to these droplets are detected after exposure, see e.g., Fig. 7d. Therefore, the NaCl-induced atmospheric corrosion processes investigated in this work occur in the presence of small amounts of aqueous solution.

According to [45] and [46] the equilibrium composition of the solution in the experimental conditions is 8.0% NaCl by weight. The corresponding solution does not form ice at −4 °C [47]. Mg–Al alloys are protected by a partially protective surface film that is dominated by brucite, $Mg(OH)_2$[18] and [19]. According to Godard et al.[1] and

Song et al. [11], Mg–Al alloys immersed in NaCl solution corrode by an electrochemical process where the anodic dissolution of magnesium occurs according to Eq. (4), and the cathodic reduction of water occurs at different sites (Eq. (5)). Accordingly, the dominant corrosion product is reported to be brucite, $Mg(OH)_2$, based on Eq. (6); equation(4)

$$Mg(s) \rightarrow Mg^{2+}(aq) + 2e^- \tag{5}$$

$$2H_2O + 2e^- \rightarrow H_2(g) + 2OH^-(aq) \tag{6}$$

$$Mg^{2+}(aq) + 2OH^-(aq) \rightarrow Mg(OH)_2(s) \tag{7}$$

Electrochemical corrosion in a NaCl (aq) solution is expected to result in the accumulation of $MgCl_2$ (aq) at the anodic sites and of NaOH (aq) at the cathodic sites. Because $Mg(OH)_2$ is insoluble at high pH, the film is stabilized in the cathodic areas. In contrast, the anodic dissolution of Mg in neutral solution is reported to be enhanced by Cl^- (aq) [48]. Hence, Mg corrosion is expected to be strongly localized in the presence of NaCl (aq). Brucite is an electronic insulator and therefore a poor electrode for the cathodic process. Hence, the corrosion of Mg is greatly accelerated by cathodically active, intermetallic precipitates containing, e.g., Fe [18]. Therefore, commercial Mg alloys are alloyed with Mn to decrease the detrimental effects of traces of Fe in the alloy [18] and [19]. In the presence of Mn, Fe tends to dissolve in the Al_8Mn_5 (-phase), which is not very active as a cathode. The observation that the NaCl-induced atmospheric corrosion of alloy AM50 is slower than for 99.97% Mg can be partially attributed to the beneficial effect of Mn alloying. According to Jönsson et al. [33] who investigated the atmospheric corrosion of commercial Mg–Al alloys, micro-galvanic elements are established where the anodic dissolution mainly occurs in the middle of the primary -dendrite grains and the cathodic process mainly occurs on the eutectic / constituent.

The present results are in accordance with an electrochemical nature of corrosion, as illustrated both by the tendency of corrosion to be localized and by the redistribution of sodium and chloride ions on the surface. Therefore, in CO_2-free conditions, part of the surface is covered by voluminous corrosion products while the rest is only slightly corroded (Fig. 7c and Fig. 8c and d). Also, the quantitative EDX analyses show that sodium is enriched on the uncorroded metal surface at a distance from the location of corrosion attack, indicating the presence of cathodic sites, see Fig. 12. The observation that sodium and chlorine accumulate at sites corresponding to cathodic and anodic sites, respectively, is in accordance with a recent study Song et al. [49], where Mg alloys were subjected to salt-spray and immersion exposures. In contrast, Liao et al. [24] reports that very little chlorine was detected in the corrosion pits formed on two AZ31B magnesium alloys exposed in outdoor urban atmospheric environments. It is suggested that the scarce evidence for chloride in the corrosion pits of field exposed Mg alloys is due to a relatively low exposure to chloride in combination with the leaching of the corrosion products by rain. In the present work, the availability of chloride is relatively high and there is no leaching effect at all. We should also keep in mind that the use of de-icing salt on the roads creates a corrosive environment in the wintertime that is far richer in chloride than test sites with typical urban or industrial atmospheric conditions.

In this study we calculated the stoichiometric ratio (corrosion product mass/metal loss) of the corrosion products from the gravimetric results (Table 2 and Table 3) and compared them to the corresponding stoichiometric ratios (molecular mass/ metal ion mass) of the identified corrosion products (Table 4). When evaluating the ratios presented in Table 3, it is seen that the corrosion product ratios for 99.97% Mg in the presence of NaCl and in the absence of CO_2 are about 2.4, corresponding to the stoichiometric ratio of brucite. This is in accordance with the XRD analysis which shows brucite to be the only crystalline corrosion product in this environment (see Table 5). Table 3 shows that the corrosion product ratio for alloy AM50 in the absence of salt tends to be slightly higher than 2.4. This is connected to the formation of the magnesium aluminum hydroxide meixnerite in that case which has a relatively high corrosion product ratio. Table 2 and Table 3 show significantly higher corrosion product ratios in the presence of CO_2 formed on both materials. This also agrees well with the XRD analysis that identified three different magnesium hydroxy carbonates with formulas of $Mg_5(CO_3)_4(OH)_2 \times 4H_2O$, $Mg_5(CO_3)_4(OH)_2 \times 5H_2O$ and $Mg_5(CO_3)_4(OH)_2 \times 8H_2O$ having corrosion product ratios of 3.85, 4.00 and 4.44 respectively. The lower corrosion product ratios registered for 99.97% Mg compared to AM50 in the presence of CO_2 are in accordance with the identification of brucite by XRD (Table 5).

Significantly, the average corrosion rate is much higher in the absence of CO_2 (Table 2 and Table 3). The inhibitive effect of CO_2 on the NaCl-induced atmospheric corrosion of 99.97% Mg and alloy AM50 is a relatively well-known phenomenon [29] and [50].

CO_2 neutralizes the catholyte and causes precipitation of sparingly soluble, carbonate-containing corrosion products. The formation of these corrosion products is an indication of the active role of CO_2 in the atmospheric corrosion. When CO_2 dissolves in water, it forms carbonic acid (H_2CO_3 (aq)). In alkaline conditions carbonic acid forms carbonate according to the Eqs. (7) and (8). CO_2 also reacts with brucite, which is the primary corrosion product, and forms magnesium hydroxy carbonates, see e.g., Eq. (9): equation(7)

$$H_2CO_3(aq)+OH^-(aq)\rightarrow HCO_3^-\ (aq) \tag{7}$$

$$HCO_3^-\ (aq)+OH^{-}(aq)\rightarrow CO_3^{2-}\ (aq) \tag{8}$$

$$5Mg(OH)_2(s)+4CO_2(aq)\rightarrow Mg_5(CO_3)_4(OH)_2\cdot4H_2O(s) \tag{9}$$

Accordingly, several crystalline magnesium hydroxy carbonates were identified after exposure at 4 and 22 °C. These products are non-conducting and cannot serve as cathodes. They are also less soluble than brucite at neutral pH [50] and may thus slow down corrosion by physically blocking the anodic sites. In the case of alloy AM50, CO_2 may also protect against corrosion by stabilizing the alumina component of the passive film which tends to dissolve at high pH [29]. The effect of CO_2 may be important for understanding the corrosion behavior of Mg alloys in cases where there is a limited supply of CO_2 e.g., in crevices, lap joints and beneath coatings.

Corrosion attack is more evenly distributed in the presence of CO_2 (compare Fig. 8a, b with c, d). Thus, in the presence of 400 ppm CO_2, there are no extensive uncorroded regions, corrosion initially being confined to the NaCl (aq) droplets. In recent paper by Shahabi et al. [29] reporting on the effect of CO_2 on the atmospheric corrosion of

Mg and MgAl alloys, it was suggested that in the absence of CO_2, a high pH develops in the NaCl (aq) droplet periphery which decreases surface tension at the electrolyte/oxide interface and causes spreading of the electrolyte and droplet coalescence so that large areas on the surface become electrochemically connected. It was argued that CO_2 counteracts this effect by neutralization, explaining the smaller corrosion cells in the presence of CO_2.

Fig. 13 visualizes the cross section of the alloy presented in Fig. 11. Here, attention is drawn to the anodic pits developed in the presence of CO_2 that tend to occur in the periphery of the NaCl (aq) droplets (as presented in Fig. 10 and Fig. 11) and to the magnesium containing corrosion product precipitates (magnesium hydroxy carbonate) that cover the surface of the former droplets. It is suggested that the presence of a solid precipitate on the surface of the droplet restricts the availability of CO_2 in the droplet interior. Under those circumstances one may expect high pH to develop on the cathodic areas (presumed to be mainly -phase areas) in the center of the droplet where CO_2 availability is expected to be the least. This would stabilize the brucite-based film, and the anodic reaction would run preferentially where the CO_2 availability is greater, and pH lower, i.e., in the droplet periphery.

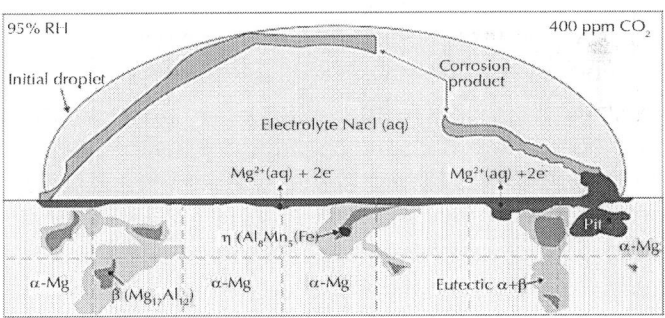

Figure 13: Schematic illustration of the cross section of the alloy AM50 in the presence of CO_2 showing microstructural constituents and the formation of the pits (corresponds to the FIB-prepared cross section shown in Fig. 11).

Fig. 13 visualizes the cross section of the alloy presented in Fig. 11. Here, attention is drawn to the anodic pits developed in the presence of CO_2 that tend to occur in the periphery of the NaCl (aq) droplets (as presented in Fig. 10 and Fig. 11) and that magnesium containing corrosion product precipitates (magnesium hydroxy carbonate) cover the surface of the former droplets. It is suggested that the presence of a solid precipitate on the surface of the droplet restricts the availability of CO_2 in the droplet interior. Under those circumstances one may expect high pH to develop on the cathodic areas (presumed to be mainly -phase areas) in the center of the droplet where CO_2 availability is expected to be the least. This would stabilize the brucite-based film, and the anodic reaction would run preferentially where the CO_2 availability is greater, and pH lower, i.e., in the droplet periphery.

The Effect of Temperature

It is well-known that variations in temperature profoundly influence the rate of atmospheric corrosion. The effect is mainly due to concomitant changes in the relative humidity that affect the amount of water on the metal surface and thereby the availability of electrolyte. In order to investigate the temperature effect *per se*, all exposures in the present study are performed at the same relative humidity (=95%). Because NaCl absorbs water to form an aqueous solution at RH > 75% in the temperature range studied, this means that the samples are exposed to NaCl (aq) solution during the exposures. It may be noted that the composition of the NaCl (aq) electrolyte (a_w = 0.95) has a very weak dependence on temperature [51]. Therefore, the amount and composition of the NaCl (aq) solution is nearly the same at the three temperatures investigated (22, 4 and −4 °C). Hence, we can be confident that the changes in corrosion behavior that we observe as a function of temperature are due to temperature only and not to differences in the availability of aqueous electrolyte on the surface or to the composition of that electrolyte. Once the influence of temperature on atmospheric corrosion is known, it becomes meaningful and interesting to investigate how variations in the relative humidity affects corrosion at different temperatures. However, such a study is outside the scope of this paper.

The introduction mentions a few laboratory investigations of atmospheric corrosion below room temperature. It is notable that the reported correlations between corrosion rate and temperature (keeping the activity of water constant) are very different. A positive correlation between corrosion rate and temperature (AZ91/NaCl and Al/NaCl) suggests that there is a rate-determining step in the corrosion process which is thermally activated, showing an Arrhenius-type behavior. This kind of behavior is expected since the chemical and electrochemical reactions involved in the anodic and cathodic reactions are thermally activated. Also, the conductivity of the electrolyte is related to ionic diffusion which is thermally activated. The inverse correlation of corrosion rate with temperature that has been reported in some cases is more unexpected [38] and [52]. In the case of the Zn/air + SO_2 system the effect was attributed to the precipitation of zinc hydroxy sulfate, forming a protective layer and removing ions from the electrolyte. The precipitate reportedly only nucleated at room temperature. Thus, $ZnSO_4$ (aq) electrolyte coexisted with solid ZnO (s) on the metal surface at low temperature, explaining the faster corrosion. In the case of Pb/air + gaseous acetic acid, it was suggested that the rate of corrosion was limited by the adsorption of acetic acid molecules on the lead surface. The adsorption of a gas on a surface is exothermic and the equilibrium amount of adsorbed gas thus decreases with increasing temperature.

The present results show a strong positive temperature dependence for the NaCl-induced corrosion of alloy AM50, especially in the presence of CO_2 (see Fig. 4b, Fig. 8a and b, Table 3). In the case of 99.97% Mg the effect of temperature is more complex (see Fig. 4a, Table 2). Thus, there is no significant correlation between temperature and corrosion rate in the absence of CO_2 while the corrosion rate is at its maximum at 4 °C in the presence of 400 ppm CO_2. It is suggested that the peculiar temperature dependence of the corrosion rate of 99.97% Mg in the presence of CO_2 is due to competing effects. Thus, the decreasing solubility of CO_2 in water with increasing temperature is expected to slow down carbonatization at higher temperature. Conversely, the hydration of CO_2 to form carbonic acid: equation(10)

$$CO_2(aq)+H_2O \rightarrow H_2CO_3(aq) \tag{10}$$

This is a relatively slow, activated process that becomes faster at high temperature, promoting carbonatization. The absence of a clear temperature dependence of the corrosion of Mg and the strong positive temperature effect reported for NaCl-induced atmospheric corrosion of Al [37] suggests that the slowing of the rate of corrosion of alloy AM50 at 4 and −4 °C is related to the aluminum content in the alloy. Hence, the strong positive temperature dependence of the corrosion rate of alloy AM50 is suggested to be due to an activated process involving aluminum. One hypothesis is that the activated process consists of the dissolution of alumina in the film formed on the surface of the alloy.

It is observed that the inhibitive effect of CO_2 on the atmospheric corrosion becomes more pronounced at lower temperatures. At −4 °C the corrosion of alloy AM50 is ~13 times slower in the presence of CO_2 compared CO_2-free condition, while the corresponding factor is about 5 at 22 °C. This can be attributed to the fact that the solubility of CO_2 (as a corrosion inhibitor) in the aqueous electrolyte increases with decreasing temperature.

As expected, the corrosion product composition is strongly affected by CO_2. The XRD diffraction patterns acquired from alloy AM50 in the absence of CO_2 and in the presence of NaCl showed mainly brucite and traces of meixnerite ($Mg_6Al_2(OH)_{18} \times 4.5H_2O$), while the corresponding exposures in the presence of 400 ppm CO_2 at 4 and 22 °C resulted in the formation of magnesium hydroxy carbonates; giorgiosite, and two types of hydromagnesite, see Table 5 and Fig. 5a and b (the corrosion product composition at −4 °C is discussed below). The magnesium hydroxy carbonates may form either by reaction of brucite with gaseous CO_2 and H_2O or by precipitation from solution.

The prevalence of carbonate-containing corrosion products after exposure to 400 ppm CO_2 is similar to the behavior of Mg alloys exposed in the ambient atmosphere, where magnesium hydroxy carbonates (notably hydromagnesite) are usually dominant [23] and [24]. Sulfates have been reported to form in outdoor environments with high SO_2 concentration [25] while brucite has been reported to form in some cases of high corrosion rates in marine environments [53]. To our knowledge, meixnerite have not been reported as corrosion products previously. The suppression of meixnerite formation by CO_2 is explained by neutralization of the catholyte by CO_2 [38]. Thus, the alumina (AlOOH or $Al(OH)_3$) component in the film becomes soluble at high pH.

Surprisingly, temperature has a major effect on the corrosion product composition on both materials in the presence of CO_2. Thus, while several crystalline magnesium hydroxy carbonates were identified at 4 and 22 °C, the same compounds are absent at −4 °C. The large amount of corrosion products detected on 99.97% Mg in 400 ppm CO_2 at −4 °C (Table 5), shows that the lack of evidence for these compounds is not due to poor sensitivity of the XRD analysis. Also, the measurements were repeated several times. The analysis of the corrosion products by FTIR (Fig. 5d) indicates that the chemical composition of the corrosion product formed in the presence of CO_2 at −4 °C is similar to that formed in the presence of CO_2 at higher temperatures. Also, the corrosion product ratio is essentially the same at the different exposure temperatures (Table 3). Thus, it can be concluded that while magnesium hydroxy carbonates dominate in the presence of 400 ppm CO_2 regardless of temperature, they tend to be non-crystalline at −4 °C and crystalline at 4 and 22 °C. In addition to the non-crystalline products, there were weak diffraction lines from an unknown crystalline compound on 99.97% Mg in 400 ppm CO_2 (designated U in Table 5). It is possible that this compound contains carbonate. It is unlikely that the prevalence of non-crystalline corrosion products in 400 ppm CO_2 at −4 °C can explain the effect of temperature on the rate of atmospheric corrosion of alloy AM50. This is because 99.97% Mg, which does not show a strong effect of temperature on corrosion rate, also does not form the crystalline hydroxy carbonates detected at higher temperature under these conditions. In addition, the temperature effect on AM50 is also evident at 4 °C where the magnesium hydroxy carbonates are crystalline.

Yet another set of diffraction lines corresponding to an unknown phase were detected after exposure at −4 °C (designated U1 in Table 5). It is suggested that U1 may be a hitherto unknown low-temperature form of magnesium hydroxide. Additionally, from the obtained results it can be concluded that this phase does not contain Al, carbonate or chloride as U1 occurs both on AM50 and on 99.97% Mg and both in the presence and in the absence of CO_2 and NaCl. The task of identification and characterization of the unknown compounds U and U1 is outside the scope of this paper but will be the subject of further investigation.

CONCLUSIONS

Designing new exposure equipment enabled us to investigate the effect of temperature on the NaCl-induced atmospheric corrosion of

99.97% Mg and the Mg–Al alloy AM50. The effects of exposure time and CO_2 were studied at three temperatures, 22, 4 and −4 °C. The main conclusions drawn from this study are as follows:

- The NaCl-induced atmospheric corrosion of alloy AM50 shows a strong positive correlation with temperature. The effect of temperature was especially strong in the presence of CO_2. In contrast, 99.97% Mg did not exhibit a strong temperature dependence.

- 99.97% Mg and alloy AM50 corrode in humid air at −4 °C in the presence of NaCl because of the presence of an aqueous electrolyte.

- The temperature dependence of the corrosion of alloy AM50 is tentatively attributed to its aluminum content and the increased inhibitive effect of CO_2 at low temperature is suggested to be due to an increased solubility of CO_2 in the aqueous electrolyte at low temperature.

- In the absence of CO_2 the NaCl-induced atmospheric corrosion resulted in the same crystalline corrosion products at all three temperatures. Thus, brucite was formed on both materials. On alloy AM50, meixnerite $(Mg_6Al_2(OH)_{18} \times 4.5H_2O)$ was also identified.

- NaCl-induced atmospheric corrosion in the presence of CO_2 resulted in the formation of three different magnesium hydroxy carbonates $(Mg_5(CO_3)_4(OH)_2 \times xH_2O)$ at 22 and 4 °C. At −4 °C, however, NaCl-induced atmospheric corrosion resulted in the formation of a non-crystalline carbonate-containing corrosion product and an unknown crystalline compound.

- Anodic and cathodic sites were identified by EDX analyses on the FIB-prepared cross sections. Chloride mainly accumulates in the pits while sodium accumulate at a distance from the areas of corrosion attack, indicating the cathodic sites.

- For all exposure conditions the rate of atmospheric corrosion showed a positive correlation with the amount of NaCl added. The corrosion rates in the presence of CO_2 were significantly less than in the CO_2-free exposures for both materials, regardless of temperature.

ACKNOWLEDGMENTS

The authors would like to express their thanks to The Swedish Foundation for Strategic Research (SSF) for funding this project.

REFERENCES

1.   H.P. Godard, W.B. Jepson, M.R. Bothwell, R.L. Lane The Corrosion of Light Metals Wiley and Sons, New York (1967)

2.   G.L. Markar, J. Kruger Corrosion of magnesium Int. Mater. Rev., 38 (1993), pp. 138–153

3.   K.U. Kainer, R.L. Edgar Global Overview on Demand and Applications for Magnesium Alloy Wiley-VCH Verlag GmbH (2000)

4.   K.U. Kainer, F. Kaiser Magnesium Alloys and Technology Wiley-VCH GmbH, Weinheim (2003)

5.   Z.M. Shi, G.L. Song, A. Atrens Influence of anodising current on the corrosion resistance of anodised AZ91D magnesium alloy Corros. Sci., 48 (2006), pp. 1939–1959

6.   A. Pardo, M.C. Merino, A.E. Coy, R. Arrabal, F. Viejo, E. Matykina Corrosion behaviour of magnesium/aluminium alloys in 3.5 wt.% NaCl Corros. Sci., 50 (2008), pp. 823–834

7.   M.M. Advedesian, H. Baker Magnesium and Magnesium Alloys, ASM Specialty Handbook ASM International, Materials Park (1999)

8.   P.B. Srinivasan, C. Blawert, W. Dietzet, K.U. Kainer Stress corrosion cracking behaviour of a surface-modified magnesium alloy Scripta Mater., 59 (2008), pp. 43–46

9.   G.L. Makar, K. Kruger Corrosion studies of rapidly solidified magnesium alloys J. Electrochem. Soc., 137 (1990), pp. 414–421

10.   D.B. Blücher, J.E. Svensson, L.G. Johansson The influence of CO_2, $AlCl_3 \cdot 6H_2O$, $MgCl_2 \cdot 6H_2O$, Na_2SO_4 and NaCl on the atmospheric corrosion of aluminum Corros. Sci., 48 (2006), pp. 1848–1866

11.   G. Song, A. Atrens, M. Dargusch Influence of microstructure on the corrosion of die-cast AZ91D Corros. Sci., 41 (1999), pp. 249–273

12.   R. Tunold, H. Holtan, M.B. Berge The corrosion of magnesium in aqueous solution containing chloride ions Corros. Sci., 17 (1977), pp. 353–365

13.   M.C. Zhao, M. Liu, G. Song, A. Atrens Influence of pH and chloride ion concentration on the corrosion of Mg alloy ZE41 Corros. Sci., 50 (2008), pp. 3168–3178

14.   G. Wu, Y. Fan, A. Atrens, C. Zhai, W. Ding Electrochemical behavior of magnesium alloys AZ91D, AZCe2, and AZLa1 in chloride and sulfate solutions J. Appl. Electrochem., 38 (2008), pp. 251–257

15.   F. Cao, Z. Shi, J. Hofstetter, P.J. Uggowitzer, G. Song, M. Liu, A. Atrens Corrosion of ultra-high-purity Mg in 3.5% NaCl solution saturated with $Mg(OH)_2$ Corros. Sci., 75 (2013), pp. 78–99

16.   F. Cao, Z. Shi, G. Song, M. Liu, A. Atrens Corrosion behaviour in salt spray and in 3.5% NaCl solution saturated with $Mg(OH)_2$ of as-cast and solution heat-treated binary Mg-X alloys: X = Mn, Sn, Ca, Zn, Al, Zr, Si, Sr Corros. Sci., 76 (2013), pp. 60–97

17.   Z. Shi, M. Liu, A. Atrens Measurement of the corrosion rate of magnesium alloys using Tafel extrapolation Corros. Sci., 52 (2010), pp. 579–588

18.   G. Song, A. Atrens Corrosion mechanisms of magnesium alloys Adv. Eng. Mater., 1 (1999), pp. 11–33

19.   G. Song, A. Atrens Understanding magnesium corrosion – a framework for improved alloy performance Adv. Eng. Mater., 5 (2003), pp. 837–858

20.   G. Song Recent progress in corrosion and protection of magnesium alloys Adv. Eng. Mater., 7 (2005), pp. 563–586

21.   H. Matsubar, Y. Ichige, K. Fujita, H. Nishiyama, K. Hodouchi Effect of impurity Fe on corrosion behavior of AM50 and AM60 magnesium alloys Corros. Sci., 66 (2013), pp. 203–210

22.   G.R. Meira, C. Andrade, C. Alonso, I.J. Padaratz, J.C. Borba Salinity of marine aerosols in a Brazilian coastal area-Influence of wind regime Atmos. Environ., 41 (2007), pp. 8431–8441

23.   Z. Cui, X. Li, K. Xiao, C. Dong Atmospheric corrosion of field-exposed AZ31 magnesium in a tropical marine environment Corros. Sci., 76 (2013), pp. 243–256

24.   J. Liao, M. Hotta, S. Motoda, T. Shinohara Atmospheric corrosion of two field-exposed AZ31B magnesium alloys with different grain size Corros. Sci., 71 (2013), pp. 53–61

25.   L. Yang, Y. Li, Y. Wei, L. Hou, Y. Tian Atmospheric corrosion of field-exposed AZ91D Mg alloys in a polluted environment Corros. Sci., 52 (2010), pp. 2188–2196

26.   Y.G. Li, Y.H. Wei, L.F. Hou, P.J. Han Atmospheric corrosion of AM60 Mg alloys in an industrial city environment Corros. Sci., 69 (2013), pp. 67–76

27.   R. Lindström, On chemistry of atmospheric corrosion, Doctoral thesis, Department of Chemistry, Göteborg University, Göteborg, 2003.

28.   D.B. Blücher, J.E. Svensson, L.G. Johansson, M. Rohwerder, M. Stratmann Scanning kelvin probe force microscopy, a useful tool for studying atmospheric corrosion of Mg–Al alloys in situ J. Electrochem. Soc., 151 (2004), pp. B621–B626

29.   M. Shahabi, M. Esmaily, J.E. Svensson, M. Halvarsson, L. Nyborg, Y. Cao, L.G. Johansson The influence of CO_2 on NaCl-induced atmospheric corrosion of the MgAl alloy AM50 J. Electrochem. Soc., 161 (2014), pp. C277–C287

30.   N. LeBozec, M. Jonsson, D. Thierry Atmospheric corrosion of magnesium alloys: influence of temperature, relative humidity, and chloride deposition Corrosion, 60 (2004), pp. 356–361

31.   C. Lin, X. Li Role of CO_2 in the initial stage of atmospheric corrosion of AZ91 magnesium alloy in the presence of NaCl Rare Met., 25 (2006), pp. 190–196

32.   S. Feliu, A. Pardo, M.C. Merino, A.E. Coy, F. Viejo, R. Arrabal Correlation between the surface chemistry and the atmospheric corrosion of AZ31, AZ80 and AZ91D magnesium alloys Appl. Surf. Sci., 255 (2009), pp. 4102–4108

33.   M. Jönsson, D. Persson, R. Gubner The initial steps of atmospheric corrosion on magnesium alloy AZ91D J. Electrochem. Soc., 154 (2007), pp. C684–C691

34.   M. Jönsson, D. Persson, C. Leygraf Atmospheric corrosion of field-exposed magnesium alloy AZ91D Corros. Sci., 50 (2008), pp. 1406–1413

35. S. Feliu, C. Maffiotte, J.C. Galván, V. Barranco Atmospheric corrosion of magnesium alloys AZ31 and AZ61 under continuous condensation conditions Corros. Sci., 53 (2011) 1865-1827

36. R. Arrabal, E. Matykina, A. Pardo, M.C. Merino, K. Paucar, M. Mohedano, P. Casajús Corrosion behaviour of AZ91D and AM50 magnesium alloys with Nd and Gd additions in humid environments Corros. Sci., 55 (2012), pp. 351–362

37. D.B. Blücher, J.E. Svensson, L.G. Johansson The NaCl-induced atmospheric corrosion of aluminum: the influence of carbon dioxide and temperature J. Electrochem. Soc., 150 (2003), pp. B93–B98

38. A. Niklasson, L.G. Johansson, J.E. Svensson The influence of relative humidity and temperature on the acetic acid vapour-induced atmospheric corrosion of lead Corros. Sci., 50 (2008), pp. 3031–3037

39. J.E. Svensson, L.G. Johansson The temperature-dependence of the SO_2-induced atmospheric corrosion of zinc; a laboratory study Corros. Sci., 38 (1996), pp. 2225–2233

40. J. Chen, J.Q. Wang, E.H. Han, W. Ke Effect of temperature on initial corrosion of AZ91 magnesium alloy under cyclic wet-dry conditions Corr. Eng. Sci. Technol., 46 (2011), pp. 267–277

41. J. Balej Water vapour partial pressures and water activities in potassium and sodium hydroxide solutions over wide concentration and temperature ranges Int. J. Hydro. Energy, 10 (1985), pp. 233–243

42. R.M. Wang, A. Eliezer, E.M. Gutman An investigation on the microstructure of an AM50 magnesium alloy Mater. Sci. Eng., A, 355 (2003), pp. 201–207

43. C.B. Baliga, P. Tsakiropoulos Development of corrosion resistant magnesium alloys: Part 2. Structure of corrosion products on rapidly solidified Mg–16Al alloys Mater. Sci. Technol., 9 (1993), pp. 513–519

44. X. Guo, J. Chang, S. He, W. Ding, X. Wang Investigation of corrosion behaviors of Mg–6Gd–3Y–0.4Zr alloy in NaCl aqueous solutions Electrochim. Acta, 52 (2007), pp. 2570–2577

45. R.A. Robinson, R.H. Stokes Electrolyte Solutions Butterworths, London (1965)

46.   K.S. Pitzer Thermodynamics of electrolytes, II. Activity and osmotic coefficients with one or both ions univalent J. Phys. Chem., 77 (1973), pp. 268–277

47.   R.J. Bodnar Revised equation and table for determining the freezing point depression of H_2O–NaCl solutions Geochim. Cosmochim. Acta, 57 (1993), pp. 683–684

48.   G. Baril, N. Pebere The corrosion of pure magnesium in aerated and deaerated sodium sulphate solutions Corros. Sci., 43 (2001), pp. 471–484

49.   W. Song, H.J. Martin, A. Hicks, D. Seely, C.A. Walton, W.B. Lawrimore, P.T. Wang, M.F. Horstemeyer Corrosion behaviour of extruded AM30 magnesium alloy under salt-spray and immersion environments Corros. Sci., 78 (2014), pp. 353–368

50.   E. Gulbrandsen Anodic behaviour of Mg in HCO^-_3/CO_2^{-3} buffer solutions. Quasi-steady measurements Electrochim. Acta, 37 (1992), pp. 1403–1412

51.   M.J. Blandamer, J.B. Engberts, P.T. Gleeson, J.C. Reis Activity of water in aqueous systems; a frequently neglected property Chem. Soc. Rev., 34 (2005), pp. 440–458

52.   J.E. Svensson, L.G. Johansson A laboratory study of the initial stages of the atmospheric corrosion of zinc in the presence of NaCl; influence of SO_2 and NO_2 Corros. Sci., 34 (1993), pp. 721–740

53.   S. Takigawa, I. Muto, N. Hara, Corrosion in marine and saltwater environments, in: ECS Transactions, 214th ECS Meeting, 2009, pp. 71–80.

Chapter 3

Mechanistic Studies of Corrosion Product Flaking on Copper and Copper-Based Alloys in Marine Environments

Xian Zhang, Inger Odnevall Wallinder, Christofer Leygraf*

KTH Royal Institute of Technology, Div. Surface and Corrosion Science, School of Chemical Science and Engineering, Dr. Kristinas v. 51, SE-100 44 Stockholm, Sweden

ABSTRACT

The mechanism of corrosion product flaking on bare copper sheet and three copper-based alloys in chloride rich environments has been explored through field and laboratory exposures. The tendency for flaking is much more pronounced on Cu and Cu–4 wt%Sn than on Cu–15 wt%Zn and Cu–5 wt%Al–5 wt%Zn. This difference is explained by the initial formation of zinc and zinc–aluminum hydroxycarbonates on Cu15Zn and Cu5Al5Zn, which delays the formation of CuCl, a precursor of $Cu_2(OH)_3Cl$. As a result, the observed volume expansion during transformation of CuCl to $Cu_2(OH)_3Cl$ and concomitant corrosion product flaking, is less severe on Cu15Zn and Cu5Al5Zn than on Cu and Cu4Sn.

INTRODUCTION

Copper and copper-based alloys form a large group of important construction materials for outdoor applications due to their appealing visual appearance, desirable mechanical and physical properties as well as their inherent resistance to atmospheric corrosion. Significant knowledge exists in the scientific literature on patina formation on copper in marine exposure conditions, and to some extent also for some copper-based alloys in chloride rich environments [1], [2], [3], [4], [5], [6] and [7]. In sheltered marine exposure conditions cuprite (Cu_2O) is the initial phase in the evolution of the copper patina. Interaction with chlorides results in the formation of nantokite (CuCl) which commonly transforms to atacamite or the isomorphous phase paratacamite $(Cu_2(OH)_3Cl)$ as the end corrosion products [1]. These constituents have also been identified within the patina on bare copper in unsheltered marine exposure conditions [2], and observed after exposure in laboratory conditions with humidified air and predeposited NaCl [3] and [8]. The patina constituents have also been observed within the patina on restored ancient bronzes [9], [10] and [11], and nantokite, trapped within the patina of archeological bronzes, and paratacamite are important constituents in so-called bronze disease [10] and [11]. Tin oxide (SnO_2) has also been identified within the patina on bronze exposed to different environments [12] and [13].

On brass (Cu–Zn alloys) amorphous zinc hydroxycarbonate, hydrozincite $(Zn_5(CO_3)_2(OH)_6)$ and zinc oxide (ZnO) were observed to cover a large part of the surface after exposure in sites with low chloride deposition rates [6], while zinc oxide was formed in laboratory conditions with only humidified air [14]. The patina composition on Cu5Al5Zn has rarely been investigated. However, hydrozincite $(Zn_5(CO_3)_2(OH)_6)$, simonkolleite $(Zn_5Cl_2(OH)_8 \cdot H_2O)$, $Zn_2Al(OH)_6Cl \cdot 2H_2O$ and hydrotalcite $(Zn_6Al_2(OH)_{16}CO_3 \cdot 4H_2O)$ have been observed as possible patina constituents [15]. These phases have also been detected as constituents in the patina evolution on zinc and zinc–aluminum alloys in different atmospheric exposure conditions [16], [17] and [18].

In a recent study [15], the metal release and corrosion properties were investigated of copper and three commercial copper-based alloys (bronze, Cu4Sn; brass, Cu15Zn and Cu5Al5Zn) used in outdoor building applications during long-term exposures in marine atmospheric environments with four different levels of chloride deposition rates. Up to 3 years of exposure, severe corrosion product flaking was observed on Cu and Cu4Sn, but to a much lesser extent on Cu15Zn and Cu5Al5Zn. Flaking of corrosion products formed on Cu-based alloys in marine environments has occasionally been reported earlier [19], but the underlying mechanism seems largely unexplored.

A more detailed study was therefore undertaken to possibly reveal the mechanism of corrosion product flaking on copper-based alloys in marine environments. The starting point was the possible importance of nantokite for corrosion product flaking [15]. Complementary laboratory exposures were conducted with synthetic nantokite predeposited on copper, and its transformation to paratacamite was studied in detail during well-controlled cyclic wet/dry exposures. The influence of humidity and chlorides on the corrosion product composition, morphology and degree of flaking was investigated in situ with IRAS (Infrared Reflection Absorption Spectroscopy) and ex *situ* with CRM (Confocal Raman Microspectroscopy), SEM/EDS (Scanning electron microscopy/Energy dispersive X-ray analysis) and GIXRD (Grazing incidence X-ray diffraction).

MATERIALS AND METHODS

Materials and Surface Preparation

Commercial surfaces of bare copper sheet (purity 99.98%), Cu4Sn (96 wt% Cu and 4 wt% Sn), Cu15Zn (85 wt% Cu and 15 wt% Zn) and Cu5Al5Zn (89 wt% Cu, 5 wt% Al, 5 wt% Zn, and 1% others) were provided via the European Copper Institute, Brussels. For more detailed information on bulk alloy composition, see [15].

The bare samples were cut to a dimension of 1 × 1 cm for in situ IRAS exposure and 2 × 5 cm for climatic chamber exposure. Each sample was mechanically abraded on 800# and 2400# SiC paper and then diamond polished down to 1 μm to obtain a fresh surface. All samples were cleaned ultrasonically in analytical grade ethanol for 10 min and dried by cold nitrogen gas then stored in a desiccator overnight.

A nantokite layer (CuCl) was grown artificially on the bare copper surface. The synthesis protocol of artificial nantokite followed the procedure in [20] and [21]. After the same surface preparation as above, bare copper samples were immersed for 1 h in a saturated $CuCl_2·2H_2O$ solution (>78 g of $CuCl_2·2H_2O$ per 100 mL of deionized water). The samples were then rinsed with deionized water followed by immediate drying in N_2 and storage in a desiccator overnight.

Laboratory Wet/Dry Cycle Exposure Conditions

Parallel experiments were conducted by means of in situ IRAS (Infrared reflection absorption spectroscopy) and climatic chamber exposures. Wet/dry cycle experiments were carried out on copper and copper-based alloys samples exposed to the following cyclic exposure conditions: NaCl pre-deposition (4 μg $NaCl/cm^2$), the first cycle 4 h (RH 90%) and 2 h (RH 0%), the second cycle 16 h (RH 90%) and 2 h (RH 0%). These cycles were repeated several times. Detailed information of the wet/dry experiments is given elsewhere [17].

Prior to exposure, NaCl (in a saturated 99.5% ethanol solution) was applied onto the surfaces by means of a transfer pipette. Upon ethanol evaporation, NaCl crystals were relatively homogeneously distributed over the surface. The amount of deposited NaCl was weighed by means of a microbalance (Mettler Toledo Excellence) and normalized to the geometric surface area of the samples. Detailed information on the procedure for pre-deposition of NaCl is given elsewhere [22].

Climatic chamber exposure was conducted by using WEISS WK1000 climatic chamber. Four different samples (Cu, Cu4Sn, Cu15Zn and Cu5Al5Zn) were attached on each Plexiglas fixture and exposed 45° from the horizontal in the chamber. The exposure of all samples on Plexiglas fixtures started simultaneously following the wet/dry cycles described above. After 1, 2, 6 and 14 cycles (corresponding to 6 h, 1, 3, 7 days) each Plexiglas fixture was withdrawn. Further ex situ surface characterization was performed on exposed samples.

In situ IRAS analysis of samples with predeposited NaCl was performed in a chamber inside the infrared spectrometer with humidified air following the same wet/dry cycles as during the climatic chamber exposure. By mixing dry and wet pre-cleaned compressed air of reduced CO_2 (lower than 20 ppm), controlled humidified conditions were obtained. In order to obtain ambient CO_2 concentration, 350 ppm, a small flow of air with 1.17% CO_2 from a CO_2 cylinder was added into the humidity chamber. Experimental details are given elsewhere [23]. In situ IRAS spectra of corrosion product formation were generated during the dry cycles.

Long-Term Field Exposure Conditions

The same batch of non-treated copper and copper-based alloys as investigated in the laboratory was also exposed in unsheltered conditions (45° from the horizontal, facing south) at the marine site of Brest, France, for 3 years. This site corresponds to site 1 in the previous study [15] and is located close to the coastal line (<5 m, on the seashore), where the measured deposition rate of chlorides was categorized as S_3 (300–1500 mg m^{-2} d^{-1}).

Exposed samples for cross-sectional analysis were embedded in a conductive polymer followed by polishing with 0.25 µm diamond paste and by subsequent polishing for 15 min using OP-S suspension (Struers A/S, Denmark) and water, whereby a near-mirror like surface of each cross-section was obtained.

Surface and Patina Analysis

IRAS (Infrared reflection absorption spectroscopy) measurements were employed to identify functional groups within the patina formed. The spectra were recorded by means of a commercial Digilab 4.0 Pro FTIR (Fourier transform infrared reflection spectroscopy) spectrometer with a MCT detector (4000–400 cm^{-1}), acquiring 1024 scans with a resolution of 4 cm^{-1} for each spectrum. The results are presented in absorbance units ($-\log (R/R_0)$), where R is the reflectance of the exposed sample surface and R_0 the reflectance of the non-exposed sample [24].

SEM/EDS (Scanning electron microscopy and energy dispersive spectroscopy) analysis was conducted to obtain patina morphology and elemental information. Cross-sections were analyzed using a LEO 1530 instrument with a Gemini column, upgraded to a Zeiss Supra 55 (equivalent) and an EDS X-Max SDD (Silicon Drift Detector) 50 mm^2 detector from Oxford Instruments. The images of cross-sections were recorded by using a backscattered electron (BSE) detector with an accelerating voltage of 15 kV. Surface analysis was performed using a FEI-XL 30 Series instrument, equipped with an EDAX Phoenix EDS system with an ultra-thin windows Si–Li detector. All surface images (75% SE, secondary electrons and 25% BSE, backscattered electrons) were obtained using an accelerating voltage of 20 kV.

CRM (Confocal Raman microspectroscopy) measurements were carried out to display the lateral distribution of functional groups within the patina using a WITec alpha300 system equipped with laser sources of wavelength 532 nm and 785 nm. A Nikon NA0.9 NGC objective was used giving a lateral resolution of around 300–400 nm.

GIXRD (Grazing incidence X-ray diffraction) analysis was used to identify the crystalline corrosion products. The measurements were performed with an X'pert PRO PANALYTICAL system, equipped with an X-ray mirror (Mo K radiation) and a 0.27° parallel plate collimator on the diffracted side. Scanning patterns were generated from a 1 × 1 cm area at a grazing angle of 88° versus the surface.

RESULTS AND DISCUSSIONS

Evidence is first presented of the different flaking characteristics during marine field exposure of bare Cu and Cu4Sn one hand, and of Cu15Zn and Cu5Al5Zn on the other hand. From this is proposed that nantokite plays an important role for the flaking process, and the subsequent sections present more detailed laboratory and field studies on the occurrence and role of nantokite for the patina evolution, including the volume expansion during transformation of nantokite to paratacamite. Finally, the observation of zinc and zinc–aluminum hydroxycarbonates on the copper-based alloys with least flaking characteristics is presented, and possible implications for the flaking mechanism discussed.

Flaking Characteristics on Copper and Copper-Based Alloys at Field Conditions

Fig. 1 displays top views of patina morphologies, as seen with SEM, of Cu, Cu4Sn, Cu15Zn and Cu5Al5Zn after 1 year of exposure at the marine site. Poorly adherent corrosion products that have detached as flakes can be revealed on both Cu (a) and Cu4Sn (b), whereas Cu15Zn (c) and Cu5Al5Zn (d) exhibit minor or no flaking at all, respectively. Based on many observations throughout the three years of exposure, flaking was in general quite severe for Cu and Cu4Sn, while only minor degree of flaking was seen for Cu15Zn and no flaking at all for Cu5Al5Zn.

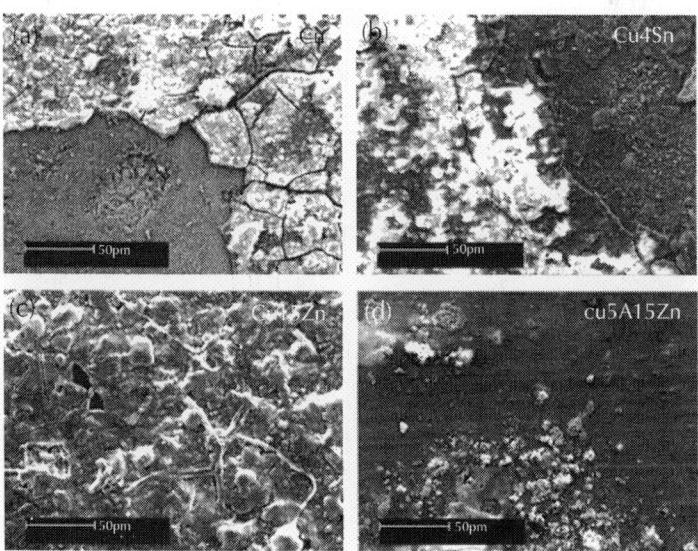

Figure 1: Top view SEM images of Cu (a) and Cu4Sn (b) surfaces with loosely adherent corrosion products (severe flaking), and Cu15Zn (c) and Cu5Al5Zn (d) surfaces with more adherent corrosion products (minor and no flaking) after 1 year of marine unsheltered field exposure.

SEM-images of cross-sections of patinas on Cu (a), Cu4Sn (b), Cu15Zn (c) and Cu5Al5Zn (d) after 2 years exposure are displayed in Fig. 2. In general, a two-layer structure of the patina was observed for all materials (a three layers structure for Cu4Sn after 3 years exposure) and the total thickness of the patina decreased as Cu4Sn > Cu ≈ Cu15Zn > Cu5Al5Zn, a trend prevailing throughout the entire three years of exposure. Further EDS line-analyses were performed on the cross-sections to reveal the corresponding elemental distribution. The main constituents within the patina layers were very similar for all copper and copper-based alloys, showing a non-uniform inner layer predominantly composed of Cu and O, and a porous outer richer in Cl, besides Cu and O.

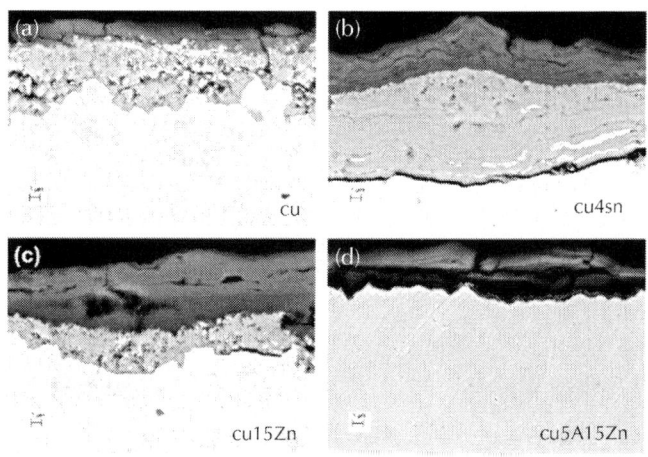

Figure 2: SEM images of cross-sections of corrosion patina formed on Cu (a), Cu4Sn (b), Cu15Zn (c), and Cu5Al5Zn (d), illustrated after 2 years of marine unsheltered field exposure.

As an example, the SEM-image of the cross-section of Cu4Sn after 3 years exposure is shown in Fig. 3a, and the corresponding elemental distribution, as viewed by EDS-mapping, is shown in Fig. 3b (Cl), c (Cu), d (O) and e (Sn) respectively. A poorly adherent outer patina layer is clearly seen with significant amounts of Cl, Cu and O, probably composed of paratacamite/atacamite, $Cu_2(OH)_3Cl$, similar to the findings from previous studies [2] and [9]. Two layers in the inner part of the patina are also seen that contain streaks of Cu + O and Sn + O respectively, probably as cuprite (Cu_2O) and tin oxide (SnO_2) respectively. Between these two inner layers there is an interfacial region enriched with Cu and Cl, probably as nantokite (CuCl). Layers of nantokite between Cu_2O and the substrate or within the outer layer have been observed with SEM/EDS after long term field and/or laboratory exposures [2], [9] and [25], and also identified with XRD[1], [2], [3], [4] and [26]. Cuprite (Cu_2O) and paratacamite (or possibly atacamite, $Cu_2(OH)_3Cl$) were confirmed by means of GIXRD also in this study. From now on this copper hydroxychloride will only be denoted paratacamite herein, although it may consist of atacamite as well.

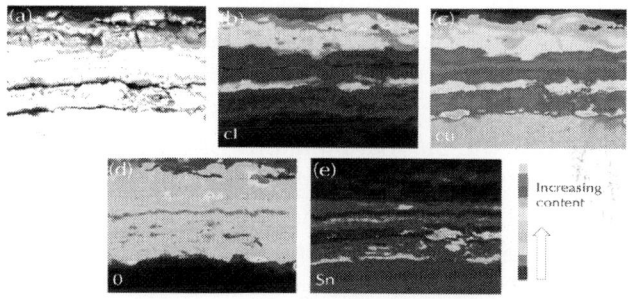

Figure 3: SEM image (a) and elemental distribution of chlorine (b), copper (c), oxygen (d) and tin (e) within the cross-section of patina formed on Cu4Sn, seen after 3 years of marine unsheltered field exposure.

In summary, after long-term marine field exposure heterogeneous and loosely adherent patina layers were observed on bare Cu sheet and Cu4Sn that easily flaked, while more adherent layers within the patina of Cu15Zn and Cu5Al5Zn were observed. As confirmed later, the outer patina layer is mainly composed of paratacamite, the inner layer mainly of cuprite, and the interfacial region in-between of nantokite. Based on these findings it is proposed that flaking is connected with the formation of nantokite. This hypothesis is further elucidated in the next sections.

Characterization and Evolution of a Synthetic Nantokite Layer on Cu

To explore the importance of nantokite as a precursor for the formation of atacamite and the underlying mechanism of flaking, an artificial nantokite layer was synthesized and applied on bare copper sheet and subsequently exposed to wet/dry cycles at 90% RH in the climatic chamber. Following the protocol of artificial nantokite described in the experimental part, the nantokite layer was synthesized by immersion of copper in saturated $CuCl_2$ solution [20]. The resulting nantokite layer appeared with varying color, ranging from white, gray, black to yellow with thickness of approximately 10 µm, as examined by a microprocessor coating thickness gauge. Fig. 4a is a SEM image of the morphology of the nantokite layer formed, which is composed of approximately 2 µm sized crystals of triangular side uniformly distributed over the copper surface, similar to previous findings [27]. Quantitative elemental analyses by means of EDS show the crystals to consist of Cu (atomic percentage 47%) and Cl (48%) and small amounts of O (5%). The GIXRD pattern of this corrosion layer is shown in Fig. 5 and displays strong similarity to the standard pattern of nantokite, besides some other weak peaks which were assigned to the copper substrate (Fig. 5, unexposed). To further confirm the existence of nantokite, the layer was also investigated with CRM. The Raman spectrum of commercial nantokite powder exhibits an intense peak at 463 cm^{-1}[28] and [29]. However, when nantokite is part of a corrosion product, it has a very poor Raman response and degrades through thermal heating, and also rapidly transforms to the basic copper hydroxide chloride at humid conditions, making it difficult to measure [26] and [30]. The Raman spectrum reported herein on artificial nantokite always showed an intense peak at 290 cm^{-1} and two less intense peaks at 613 and 1110 cm^{-1}, displayed in Fig. 4b. Two different laser wavelengths (532 nm and 785 nm) were used for identification of this layer and the same Raman spectra were recorded. This Raman spectrum of artificial nantokite is very similar to the reference spectrum of tenorite, CuO [31]. CuCl is a thermally unstable phase, which can easily degrade through dechlorination to form CuO [32] and [33]. So we suppose the oxidation of nantokite was initiated by the highly focused laser beam simultaneously when performed CRM measurements and the end product present for the Raman spectrum can be tenorite [31]. Assuming that the detection of tenorite during Raman analysis is caused by the highly focused laser beam during Raman measurements, it is therefore concluded that the synthetically produced layer on copper formed in $CuCl_2$ solution mainly consists of nantokite. This nantokite layer with its characteristic Raman spectrum is from now used as starting point to further explore the patina formation on exposed copper and copper-based alloys after exposures in laboratory conditions.

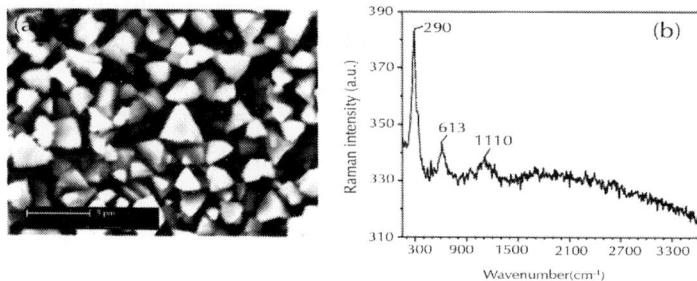

Figure 4: SEM image (a) and Raman spectrum (b) of an artificial nantokite layer on copper.

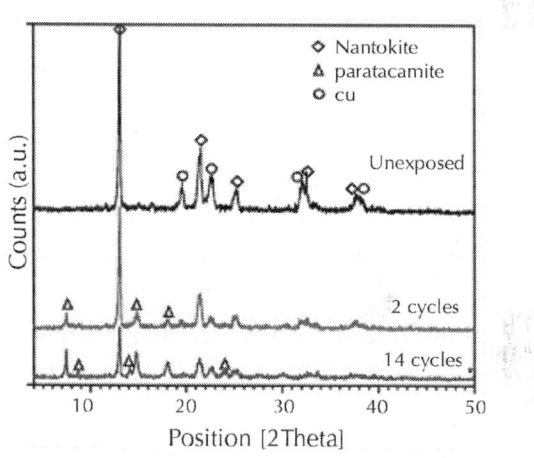

Figure 5: GIXRD diffraction patterns of the evolution of a synthetic nantokite layer on the copper surface, after 0 (unexposed), 2 and 14 wet/dry cycles at 90% RH, and with predeposited 4 $\mu g/cm^2$ NaCl.

The artificial nantokite layer on copper was next exposed in the climatic chamber to an increasing number of wet/dry cycles and analyzed with GIXRD. The visual appearance of the resulting corrosion layer changed gradually and became light greenish and powdery, whereby its thickness continuously increased to about 50 μm, examined by a microprocessor coating thickness gauge after 14 cycles exposure. In accordance with previous field exposed samples, GIXRD-analyses of the corroded layer confirmed the gradual transformation of artificial nantokite to paratacamite, see Fig. 5 and Table 1.

Table 1: Observed 2 and d-spacing values (Å) from diffraction patterns for nantokite, paratacamite and Cu as shown in Fig. 5 (Mo K radiation)

Compound	Position [°2]	d-spacing (Å)
Nantokite	13.04	3.13
	21.30	1.92
	25.16	1.63
	32.55	1.27
	37.74	1.10
Paratacamite	7.47	5.46
	8.67	4.70
	14.12	2.89
	14.82	2.76
	18.03	2.27
	24.02	1.77
Cu	19.54	2.09
	22.61	1.81
	32.11	1.28
	38.33	1.08

It is evident that the transformation of nantokite to paratacamite results in a volume expansion of the phases involved. This process may create internal physical stresses within the corrosion layer, resulting in subsequent flaking of corrosion products [10].

In summary, the artificial nantokite layer was successfully synthesized on bare copper sheet, as confirmed with SEM/EDS and GIXRD, and consists of triangular crystals with a characteristic CRM spectrum showing three peaks at 290, 613 and 1110 cm^{-1} respectively. After cyclic wet/dry exposures in the climatic chamber at 90% RH the layer transformed from nantokite to paratacamite, a process that was associated with a significant volume expansion.

The Significance of Nantokite in the Patina Formed At Field Conditions

Having established the reference Raman spectrum of nantokite, further CRM measurements were performed on the cross-sections of the patinas formed on pure Cu, Cu4Sn, Cu15Zn and Cu5Al5Zn after marine exposure for 3 years, see Fig. 6. The combined CRM images clearly distinguish the layered structure of the patina with its main constituents. Cuprite, Cu_2O (main peaks at 219, 424 and 636 cm^{-1}) [34], is the main constituent of the inner layer and paratacamite, $Cu_2(OH)_3Cl$ (main peaks at 373, 519, 930, 977, 3364 and 3447 cm^{-1}) [28] the main constituent of the outer layer of the patina on bare copper sheet and all copper-based alloys. A discontinuous thin nantokite layer is intertwined between the inner and outer layers, as evidenced from the Raman peak positions in the previous section representative of artificial nantokite, CuCl (main peaks at 290, 613 and 1110 cm^{-1}). By comparing CRM images in Fig. 6, the occurrence of nantokite was significant on Cu and Cu4Sn, less common on Cu15Zn and non-significant on Cu5Al5Zn. It is interesting to note that this order is the same as the tendency for corrosion product flaking, providing an important piece of evidence for the important role of nantokite in the flaking process.

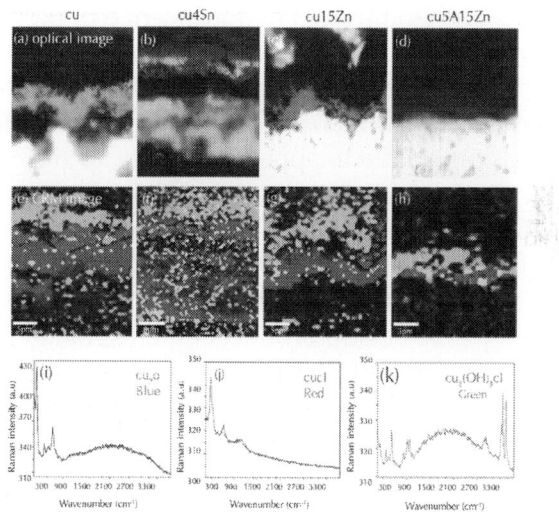

Figure 6: Optical images (a–d) and combined Raman mapping images (a laser source of 532 nm, (e–h)) obtained with CRM on Cu_2O (blue inner layer, integrated between 150 and 250 cm^{-1}, (i)), CuCl (red intertwined layer, integrated between 250 and 350 cm^{-1}, (j)) and the OH band in $Cu_2(OH)_3Cl$ (green outer layer, integrated between 3300 and 3500 cm^{-1}, (k)) of cross-sections of patina formed on Cu, Cu4Sn, Cu15Zn and Cu5Al5Zn after 3 years of marine exposure.

In summary, the patina formed on bare copper sheet and the copper-based alloys during marine long-term exposure consists of an inner layer of cuprite, an outer layer of paratacamite and a layer of nantokite in between, the occurrence of which is the same as the tendency for corrosion product flaking of the materials investigated.

Patina Evolution and Characteristics in Laboratory Exposures

In this section the formation of nantokite is elucidated further, this time by exposing all materials with predeposited NaCl (4 μg/cm^2) to the same repeated wet/dry exposures as before, and analyze the resulting corrosion products with SEM and CRM.

After exposure to 1 cycle (4 h at 90% RH followed by 2 h at 0% RH), a few ring-formed corrosion features appeared on copper and all copper-based alloys (Fig. 7a–d). These were induced by the spreading of NaCl droplets formed through deliquescence of predeposited NaCl when increasing the relative humidity from dry to 90% [17]. However, different morphologies of the ring-shaped features were observed, suggesting that the initial interaction of the droplets with the surfaces can vary. Corrosion products appearing white in the SEM images of Cu and Cu4Sn initially form along the periphery of the rings (Fig. 7a and b), whereas on Cu15Zn and Cu5Al5Zn the corrosion products preferentially form inside the much darker appearing rings (Fig. 7c and d). Compositional analysis by means of EDS showed the presence of Cu, O and Cl in the corrosion products on all surfaces and also Sn on Cu4Sn, Zn on Cu15Zn and Zn and Al on Cu5Al5Zn. White granular shaped corrosion products exhibited a higher content of Cl and O than the remaining corrosion products. After exposure to 14 cycles (in all 7 days), severe corrosion effects can be seen on Cu and Cu4Sn (Fig. 7e and f) with frequent rounded corrosion products almost covering the whole surface, independent of the original positions of NaCl droplets (Fig. 7a and b). Agglomerates of corrosion products are randomly seen on Cu and Cu4Sn, probably originating from the aggregation of white granular corrosion products. The corrosion effects seen on Cu15Zn and Cu5Al5Zn (Fig. 7g and h) after 14 cycles are much less severe and more uniform than on Cu and Cu4Sn, and the ring-like features seen after 1 cycle are still visible. Elemental analysis of corrosion product showed also here the presence of mainly Cu, O and Cl.

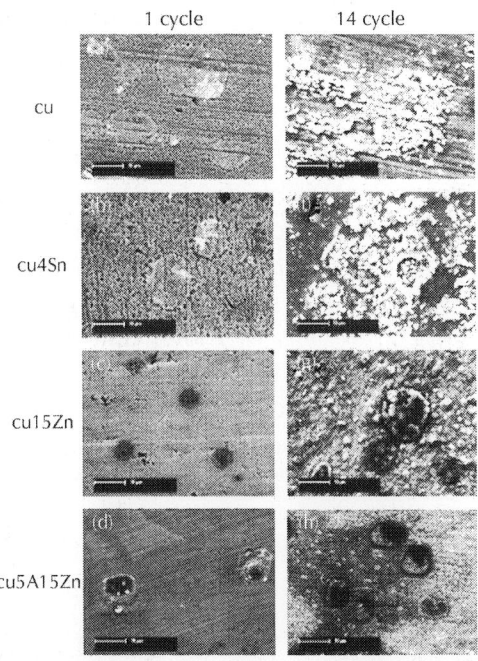

Figure 7: SEM images of patina formed on Cu, Cu4Sn, Cu15Zn and Cu5Al-5Zn with predeposited 4 µg/cm² NaCl exposed for 1 (a–d) and 14 (e–h) cycles at 90% RH.

Complementary CRM measurements of local areas (25 × 25 µm) were performed of Cu4Sn with predeposited NaCl (4 µg/cm²) after exposure to the same wet/dry cycles in the climatic chamber as before. The colors and integrated bands for different patina constituents are the same as in Fig. 6. A ring-like corrosion feature was selected after 1 cycle of exposure (Fig. 8a), and the formation of cuprite surrounding the periphery of the ring is evident (Fig. 8f). The characteristic morphology of cuprite crystals can also be seen in the corresponding SEM-image (Fig. 7b) along the periphery and also randomly distributed elsewhere. The origin of the electrochemically driven formation of cuprite has been studied in more detail elsewhere [22]. After 2 cycles not only cuprite but also nantokite can be identified in ring-like features, as seen in Fig. 8b and g. Cuprite covers most of the surface while nantokite is concentrated along and inside the periphery of the ring. In agreement with earlier studies on nantokite [10], the morphology and distribution of nantokite is very similar to the white corrosion products with small granular features seen inFig. 7b. after 6 cycles (Fig. 8c and h), cuprite and nantokite appears to cover a large part of the surface with clusters of nantokite inside circular features and cuprite outside. Nantokite has been reported to form through reaction of cuprous ions from dissolved cuprite and chloride ions [1] and [4]. After 14 cycles, the circular features are not visible any longer, only shapeless clustered features. A local area was selected and is shown in Fig. 8d. Because of volume expansion of the corrosion products the focus of the optical camera in the CRM had to be lifted by about 1 μm in order to obtain a sharp image, see Fig. 8e. Fig. 8i shows that the surrounding lower region mostly consists of cuprite, while Fig. 8j provides evidence of paratacamite in the upper region. The transformation of nantokite to paratacamite is facilitated by the porous nature of the patina allowing penetration of moisture and oxygen [10]. Evidence for the concomitant volume expansion within the patina during this transformation process was confirmed by the height difference between nantokite and paratacamite, cf. Fig. 8d–j.

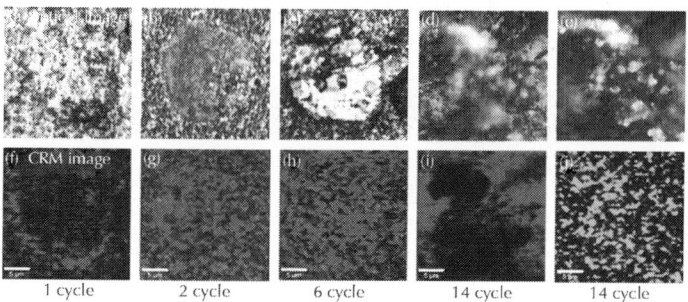

1 cycle   2 cycle   6 cycle   14 cycle   14 cycle

Figure 8: Optical images (a–e) and corresponding combined Raman mapping images (a laser source of 532 nm, (f–j)) obtained with CRM of the Cu_2O band (blue, integrated between 150 and 250 cm^{-1}), CuCl band (red, integrated between 250 and 350 cm^{-1}) and the OH band in $Cu_2(OH)_3Cl$ (green, integrated between 3300 and 3500 cm^{-1}), of the patina formed on Cu4Sn with predeposited 4 $μg/cm^2NaCl$ exposed for 1, 2,6 and 14 wet/dry cycles at 90% RH. (e) Was obtained at the same area as (d) but with the focus of the camera in the CRM lifted 1 μm.

Fig. 9 displays the patina formation on Cu, Cu15Zn and Cu5Al5Zn with predeposited NaCl after exposure to 6 wet/dry, obtained with CRM in the same way as in Fig. 8. Both cuprite and nantokite were identified on all surfaces. However, the amount of nantokite was significantly lower for Cu15Zn (Fig. 9e) and Cu5Al5Zn (Fig. 9f) than for Cu (Fig. 9d) and Cu4Sn (Fig. 8h), in agreement with findings in Section 3.3 from the field exposed samples. Furthermore, no paratacamite could be detected with CRM on Cu15Zn and Cu5Al5Zn, while on Cu and Cu4Sn this phase was easily detected.

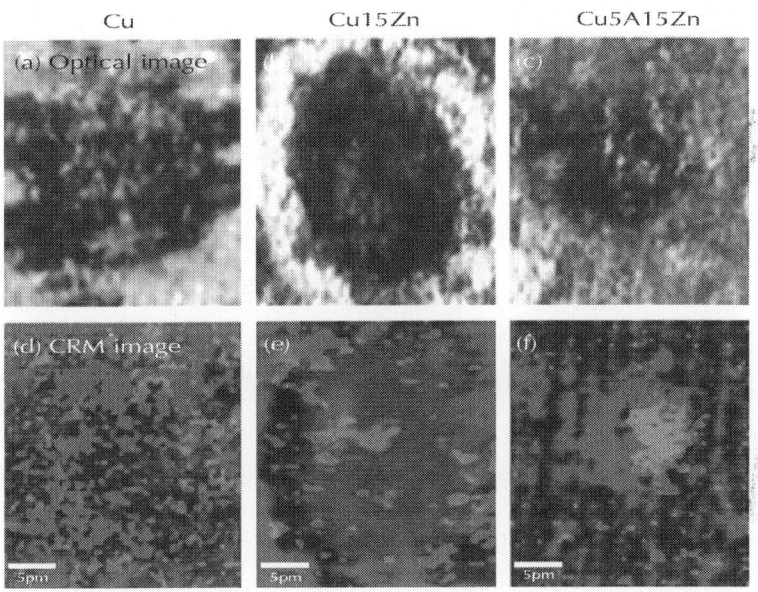

Figure 9: Optical images (a–c) and corresponding combined Raman mapping images (a laser source of 785 nm, (d–f)) obtained with CRM of the Cu_2O band (blue, integrated between 100 and 200 cm^{-1}) and CuCl band (red, integrated between 250 and 350 cm^{-1}), of corrosion patina formed on Cu, Cu15Zn and Cu5Al5Zn with predeposited 4 $\mu g/cm^2$ NaCl exposed for 6 wet/dry cycles at 90% RH.

In summary, after exposure of all samples with predeposited NaCl to an increasing number of wet/dry cycles at 90% RH of all samples with predeposited 4 $\mu g/cm^2$ NaCl, ring-like corrosion effects could be seen initially as the result deliquescence effects of NaCl. Cuprite formed initially and soon covered all surfaces. Cuprite was followed by nantokite formation, which was much more abundant on Cu and Cu4Sn than on Cu15Zn and Cu5Al5Zn, following the same trend as in marine field exposure. Paratacamite could be detected after 14 cycles on Cu and Cu4Sn, but not on Cu15Zn and Cu5Al5Zn. Comparing all results, the same reaction pathway for the patina formation was revealed in the field exposures and in the laboratory exposures, with presynthesized CuCl as well as predeposited NaCl.

The Influence of Surface Characteristics on Nantokite Formation

The previous section suggested different interactions between the NaCl droplets and the surfaces of the materials investigated. Further complementary analyses were therefore performed to examine copper and the copper-alloys surfaces with predeposited NaCl and exposed to cyclic dry/wet exposures. This time IRAS was used to analyze the surface or near-surface composition to reveal the overall corrosion product formation in an extended surface region, rather than a local. Fig. 10 exhibits the ex situ IRAS spectra in the wavenumber range from 500 up to 4000 cm^{-1} obtained after 14 wet/dry cycles for Cu, Cu4Sn, Cu15Zn and Cu5Al5Zn, respectively. Starting from lower wavenumber and going upwards the peak at 651 cm^{-1} is attributed to the vibration of Cu$_2$O, cuprite [30], formed on all surfaces. The peaks in the range from 700 to 1100 cm^{-1} may originate from the bending mode of Cu–O–H and for Cu15Zn and Cu5Al5Zn possibly also from bending modes of Zn–O–H and/or Al–O–H [30] and [35]. A broad band from 1300 to 1600 cm^{-1} with two peaks located at 1395 and 1515 cm^{-1} are clearly seen in the spectra of Cu15Zn and Cu5Al5Zn, but not for Cu and Cu4Sn. This broad band most likely originates from anti-symmetric stretching modes of carbonate (CO$_3^{2-}$) [36], and will be discussed in more detail later on. In the higher wavenumber range for Cu15Zn and Cu5Al5Zn a broad band is seen ranging from 3000 to 3700 cm^{-1} which is commonly attributed to the presence of hydroxide ions (OH$^-$) or water. In all, the results suggest the presence of hydoxycarbonates mainly on Cu15Zn and Cu5Al5Zn. It should be added that the main peaks from CuCl are located outside the investigated range of wavenumbers.

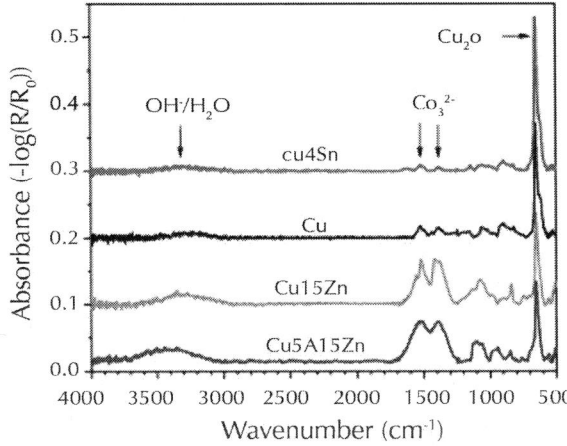

Figure 10: Ex situ IRAS spectra obtained on Cu, Cu4Sn, Cu15Zn and Cu-5Al5Zn with 4 µg/cm² predeposited NaCl after exposure to 14 wet/dry cycles at 90% RH.

The exposures were repeated to obtain an in situ monitoring with IRAS of the growth of the main peaks, with a CO_2-concentration of 350 ± 50 ppm, in order to simulate the natural atmosphere. Fig. 11 displays in situ IRAS spectra in the range from 500 up to 2000 cm^{-1} obtained on Cu (a), Cu4Sn (b), Cu15Zn (c) and Cu5Al5Zn (d), predeposited with NaCl, after exposure to 1, 2 and 6 wet/dry cycles. Cu_2O, cuprite (648 cm^{-1}) appeared immediately after the first cycle and its intensity increased with increasing number of cycles. In agreement with Fig. 10, the most pronounced feature are the peaks between 1300 and 1600 cm^{-1}, forming evidence of carbonate (CO_3^{2-}). The exact location of the peaks in this region indicate different phases. A single peak located at 1401 cm^{-1} for Cu5Al5Zn (Fig. 11d) has been assigned to hydrotalcite ($Zn_6Al_2(OH)_{16}CO_3·4H_2O$), a zinc aluminum hydroxycarbonate, which has been reported to form on Zn–Al coatings in marine environments [35] and [37]. The peaks located at 1373 and 1504 cm^{-1} (Fig. 11c) have been assigned to hydrozincite ($Zn_5(CO_3)_2(OH)_6$) [36], a phase formed in early stages of atmospheric corrosion of bare zinc and galvanized steel [16] and [38]. Furthermore, the band located at 1600 cm^{-1} is attributed to the stretching vibrations of water and hydroxyl groups. The assignments of peaks in the carbonate region of Fig. 11 up to 6 cycles then suggests that hydrozincite forms on Cu15Zn, while hydrotalcite forms on Cu5Al5Zn.

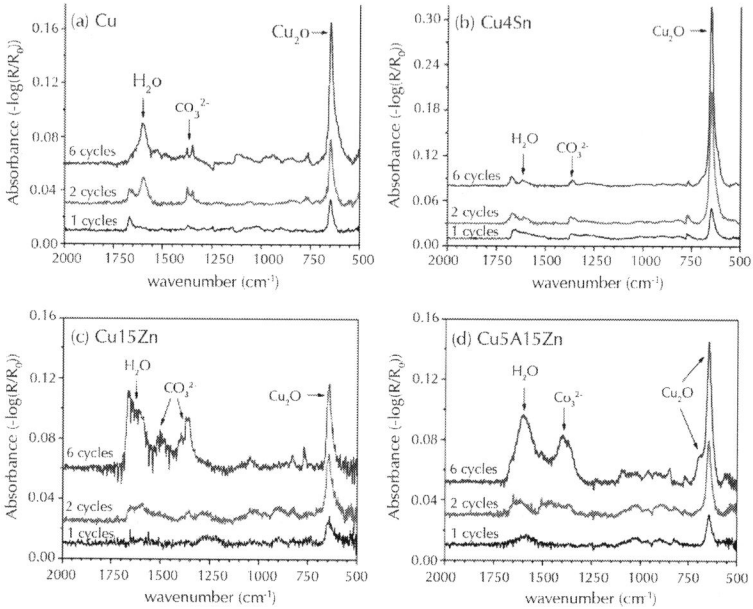

Figure 11: In situ IRAS spectra obtained on Cu (a), Cu4Sn (b), Cu15Zn (c) and Cu5Al5Zn (d) after exposure for 1, 2 and 6 wet/dry cycles at 90% RH with predeposited 4 µg/cm^2 NaCl particles and 350 ± 50 ppm CO_2.

Hydrozincite is of importance for the surface properties of bare zinc sheet and galvanized steel in chloride-rich environments [39], and hydrotalcite has a similar role for zinc–aluminum based alloys [17] and [40]. Studies by Cole et al. have shown that the atmospheric corrosion of zinc in marine environments was enhanced during the initial period of NaCl spreading, where the formation of hydrozincite was promoted in the secondary spread zone of the droplets, characterized by a higher pH. Due to its resistance to chloride diffusion and relatively high stability, hydrozincite has been shown to protect the surface from further corrosion to a much greater extent than other basic zinc compounds [41]. Hydrozincite may also possess a negative surface charge at pH lower than 7, which is attributed to the presence of conjugate base moieties on the surface [39]. A negatively charged hydrozincite surface has the capability of repelling chloride ions, which may aid in preventing chloride-induced atmospheric corrosion [39]. Similar effects may be induced by hydrotalcite, besides improved barrier properties in chloride-rich environments, as observed on Zn–

Al coatings compared to bare Zn/galvanized steel [17] and [42]. In addition to the hydroxycarbonates discussed, other phases such as ZnO, Al_2O_3 and $Zn_5(OH)_8Cl_2 \cdot H_2O$ and/or $Zn_2Al(OH)_6Cl \cdot 2H_2O$ may also enhance the barrier properties of Cu15Zn and Cu5Al5Zn in marine environments [39].

In all, the IRAS studies provide evidence of the initial formation of hydrozincite and hydrotalcite on Cu15Zn and Cu5Al5Zn, two hydroxycarbonates that can reduce the influence of chloride ions on the atmospheric corrosion of the alloys in marine environments. These findings agree well with the earlier conclusions of reduced nantokite formation on CuZn15 and Cu5Al5Zn compared to bare Cu sheet and Cu4Sn, and provide an explanation for the different behavior of the alloys investigated as discussed next.

The Mechanism of Corrosion Product Flaking On Copper and Copper-Based Alloys

An extensive study on the atmospheric corrosion of bare Cu revealed commonly occurring sequences of important patina constituents in different types of environments [1]. It starts with cuprite, is followed by nantokite and ends with copper hydroxychloride as either of two dimorphs, paratacamite (rhombohedral) or atacamite (orthorhombic). For simplicity reasons all copper hydroxychlorides have been denoted paratacamite herein.

In this study, bare copper sheet and copper-based alloys were investigated in chloride-rich environment with the aim to possibly reveal the mechanism behind corrosion product flaking. Based on current laboratory and field exposures, the earlier proposed sequence of patina evolution on bare copper was validated for the copper-based alloys as well. In addition, other phases were identified formed through reaction between the main alloying elements (Sn, Zn and Al) and the atmospheric environments. These additional phases turn out to be crucial for the ability of the corrosion products to adhere to the substrate or not.

A summary of all findings of this study are schematically displayed in Fig. 12. In moist atmosphere, cuprite (Cu_2O) forms on copper sheet and on all copper-based alloys, while zinc oxide (ZnO) and/or aluminum oxide (Al_2O_3) forms on Cu15Zn and Cu5Al5Zn. CO_2 in the atmosphere and dissolved in the aqueous adlayer rapidly forms hydrozincite $(Zn_5(CO_3)_2(OH)_6)$ and/or hydrotalcite $(Zn_6Al_2(OH)_{16}CO_3 \cdot 4H_2O)$ on Cu15Zn and Cu5Al5Zn. High chloride deposition rates result in local nantokite (CuCl) formation through reaction of chloride ions and cuprous ions. This phase turns out to be much more common on Cu and Cu4Sn than on Cu15Zn and Cu5Al5Zn, making the former two materials much more sensitive to chloride-induced atmospheric corrosion than the latter two. It easily transforms to paratacamite $(Cu_2(OH)_3Cl)$ through reaction between nantokite, water and oxygen. Paratacamite is a more voluminous corrosion product than nantokite, with a molar volume, i.e. the molar mass divided by the density, of 61.02 cm³/mol as compared to that of nantokite, 23.88 cm³/mol. Hence, the transformation of nantokite to paratacamite is connected with a significant volume expansion. The location of nantokite in the patina layer in-between cuprite and paratacamite and its continuous formation and transformation to paratacamite induces internal physical stresses, this leads to a separation between the inner patina layer (mainly cuprite) and the outer patina layer (mainly paratacamite), which induces the flaking process primarily observed on Cu and Cu4Sn.

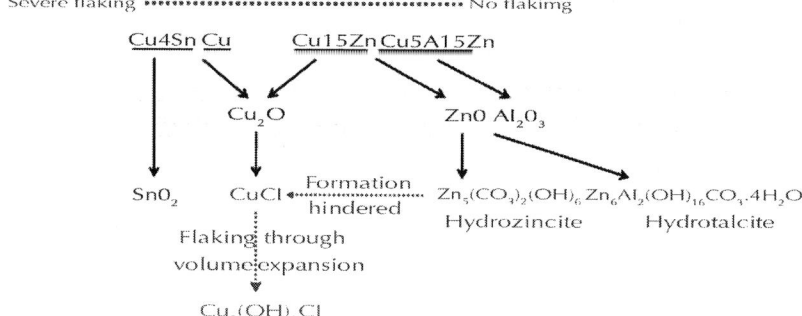

Figure 12: Sequence of phases observed in corrosion products formed on Cu, Cu4Sn, Cu15Zn and Cu5Al5Zn upon long-term unsheltered exposure in marine environments and short-term laboratory exposures in chloride-containing environments. Key processes governing the mechanism of flaking are indicated in blue.

The existence of the hydroxycarbonates hydrozincite and/or hydrotalcite improves the barrier properties of Cu15Zn and Cu5Al5Zn in chloride rich environment, through altering the surface chemistry and surface charge. Thereby the formation of nantokite and subsequent transformation to paratacamite are depressed, and then the flaking of corrosion products on these alloys is highly reduced.*

CONCLUSIONS

1. The patina on bare copper sheet and three commercial copper-based alloys, Cu4Sn, Cu15Zn and Cu5Al5Zn, formed during atmospheric exposure in a marine environment and in laboratory environment with additions of chlorides, is composed of an outer layer of $Cu_2(OH)_3Cl$ (paratacamite), an inner layer of Cu_2O (cuprite), and a layer in-between of $CuCl$ (nantokite).

2. The patina on bare Cu and Cu4Sn forms loosely adherent corrosion products that commonly flake from the surface, while that on Cu15Zn and Cu5Al5Zn forms more adherent corrosion products with significantly less tendency for spallation.

3. Artificial nantokite on bare copper sheet was successfully synthesized and appears as crystals with triangular side. After exposure to wet/dry cycles in humidified air the artificial nantokite layer transforms to a greenish voluminous patina layer of paratacamite.

4. With increasing abundance of nantokite the tendency for flaking of corrosion products increased. This is explained by the gradual transformation of nantokite to paratacamite which induces a volume expansion and internal stresses within the patina.

5. Cu15Zn and Cu5Al5Zn alloys are much less sensitive to chloride-induced corrosion product flaking than bare Cu sheet and Cu4Sn. This is attributed to the initial formation of the Zn- and Zn/Al-hydroxycarbonates hydrozincite and/or hydrotalcite on Cu15Zn and Cu5Al5Zn, which decrease the interaction of chlorides with these alloy surfaces.

ACKNOWLEDGEMENTS

Financial support from CSC, the China Scholarship Council and from ECI, the European Copper Institute, is gratefully acknowledged.

O. Karlsson at Swerea KIMAB, Sweden, and Dr. G. Herting at KTH, Sweden, are highly acknowledged for the FEG-SEM/EDS measurements on the cross-sections. Valuable help and discussions on CRM findings with Dr. B. Brandner, SP Technical Research Institute, Sweden, and general discussions with Dr. S. Goidanich, Politecnico di Milano, Italy are highly appreciated.

The grant from Nils and Dorthi Troedsson Foundation for the combined Confocal Raman AFM equipment is gratefully acknowledged.

REFERENCES

1. A. Krätschmer, I. Odnevall Wallinder, C. Leygraf, The evolution of outdoor copper patina, Corros. Sci 44 (2002) 425–450.

2. D. de la Fuente, J. Simancas, M. Morcillo, Morphological study of 16-year patinas formed on copper in a wide range of atmospheric exposures, Corros. Sci. 50 (2008) 268 285.

3. H. Strandberg, L.G. Johansson, Some aspects of the atmospheric corrosion of copper in the presence of sodium chloride, J. Electrochem. Soc. 145 (1998) 1093–1100.

4. M. Watanabe, E. Toyoda, T. Handa, T. Ichino, N. Kuwaki, Y. Higashi, T. Tanaka, Evolution of patinas on copper exposed in a suburban area, Corros. Sci. 49 (2007) 766–780.

5. L. Núñez, E. Reguera, F. Corvo, E. González, C. Vazquez, Corrosion of copper in seawater and its aerosols in a tropical island, Corros. Sci. 47 (2005) 461–484.

6. S. Goidanich, J. Brunk, G. Herting, M.A. Arenas, I. Odnevall Wallinder, Atmospheric corrosion of brass in outdoor applications: patina evolution, metal release and aesthetic appearance at urban exposure conditions, Sci. Total Environ. 412–413 (2011) 46–57.

7. J. Sandberg, I. Odnevall Wallinder, C. Leygraf, N. Le Bozec, Corrosion-induced copper runoff from naturally and pre-patinated copper in a marine environment, Corros. Sci 48 (2006) 4316–4338.

8. Z.Y. Chen, D. Persson, F. Samie, S. Zakipour, C. Leygraf, Effect of carbon dioxide on sodium chloride-induced atmospheric corrosion of copper, J. Electrochem. Soc. 152 (2005) B502–B511.

9.   M. Ghoniem, The characterization of a corroded Egyptian bronze statue and a study of the degradation phenomena, Int. J Conserv. Sci. 2 (2011) 95–108.

10. D.A. Scott, Chlorides and basic chlorides, in: D.A. Scott (Ed.), Copper and Bronze in Art – Corrosion, Colorants, Conservation, Getty Publications, Los Angeles, 2002, pp. 122–144.

11. D.A. Scott, A review of copper chlorides and related salts in bronze corrosion and as painting pigments, Stud. Conserv. 45 (2000) 39–53.

12. M. Serghini-Idrissi, M.C. Bernard, F.Z. Harrif, S. Joiret, K. Rahmouni, A. Srhiri, H. Takenouti, V. Vivier, M. Ziani, Electrochemical and spectroscopic characterizations of patinas formed on an archaeological bronze coin, Electrochim. Acta 50 (2005) 4699–4709.

13. F. Ospitali, C. Chiavari, C. Martini, E. Bernardi, F. Passarini, L. Robbiola, The characterization of Sn-based corrosion products in ancient bronzes: a Raman approach, J. Raman Spectrosc. 43 (2012) 1596–1603.

14. P. Qiu, C. Leygraf, Initial oxidation of brass induced by humidified air, Appl. Surf. Sci. 258 (2011) 1235–1241.

15. I. Odnevall Wallinder, X. Zhang, S. Goidanich, N. Le Bozec, G. Herting, C. Leygraf, Corrosion and runoff rates of Cu and three Cu-alloys in marine environments with increasing chloride deposition rate, Sci. Total Environ. 472 (2014) 681–694.

16. I. Odnevall, C. Leygraf, Reaction sequences in atmospheric corrosion of zinc, in: W.W. Kirk, H.H. Lawson (Eds.), Atmospheric Corrosion, ASTM STP 1239, American Society for Testing and Materials, Philadelphia, 1995, pp. 215–229.

17. X. Zhang, C. Leygraf, I. Odnevall Wallinder, Atmospheric corrosion of Galfan coatings on steel in chloride-rich environments, Corros. Sci. 73 (2013) 62–71.

18. X. Zhang, T.-N. Vu, P. Volovitch, C. Leygraf, K. Ogle, I. Odnevall Wallinder, The initial release of zinc and aluminum from non-treated Galvalume and the formation of corrosion products in chloride containing media, Appl. Surf. Sci. 258 (2012) 4351–4359.

19. I.R. Scholes, W.R. Jacob, Atmospheric corrosion of copper and copper-base alloys during twenty years' exposure in a marine and an industrial environment, Inst. Metals, Monogr. Rep. Ser. 34 (1970) 330–338.

20. M. Dowsett, A. Adriaens, C. Martin, L. Bouchenoire, The use of synchrotron X-rays to observe copper corrosion in real time, Anal. Chem. 84 (2012) 4866–4872.

21. A. Adriaens, M. Dowsett, G. Jones, K. Leyssens, S. Nikitenko, An in-situ X-ray absorption spectroelectrochemistry study of the response of artificial chloride corrosion layers on copper to remedial treatment, J. Anal. Atom. Spectrom. 24 (2009) 62–68.

22. Z.Y. Chen, S. Zakipour, D. Persson, C. Leygraf, Effect of sodium chloride particles on the atmospheric corrosion of pure copper, Corrosion 60 (2004) 479–491.

23. H. Gil, C. Leygraf, Quantitative in situ analysis of initial atmospheric corrosion of copper induced by acetic acid, J. Electrochem. Soc. 154 (2007) C272–C278.

24. T. Aastrup, C. Leygraf, Simultaneous infrared reflection absorption spectroscopy and quartz crystal microbalance measurements for in situ studies of the metal/atmosphere interface, J. Electrochem. Soc. 144 (1997) 2986–2990.

25. I. Constantinides, A. Adriaens, F. Adams, Surface characterization of artificial corrosion layers on copper alloy reference materials, Appl. Surf. Sci. 189 (2002) 90–101.

26. V. Hayez, V. Costa, J. Guillaume, H. Terryn, A. Hubin, Micro Raman spectroscopy used for the study of corrosion products on copper alloys: study of the chemical composition of artificial patinas used for restoration purposes, Analyst 130 (2005) 550–556.

27. J. Wang, C. Xu, G. Lv, Formation processes of CuCl and regenerated Cu crystals on bronze surfaces in neutral and acidic media, Appl. Surf. Sci. 252 (2006) 6294–6303.

28. R.L. Frost, Raman spectroscopy of selected copper minerals of significance in corrosion, Spectrochim. Acta A 59 (2003) 1195–1204.

29. R.L. Frost, P.A. Williams, J.T. Kloprogge, W. Martens, Raman spectroscopy of the copper chloride minerals nantokite,

eriochalcite and claringbullite – implications for copper corrosion, Neues Jahrb. Miner.-Monat. (2003) 433–445.

30. Z.Y. Chen, S. Zakipour, D. Persson, C. Leygraf, Combined effects of gaseous pollutants and sodium chloride particles on the atmospheric corrosion of copper, Corrosion 61 (2005) 1022–1034.

31. J.F. Xu, W. Ji, Z.X. Shen, W.S. Li, S.H. Tang, X.R. Ye, D.Z. Jia, X.Q. Xin, Raman spectra of CuO nanocrystals, J. Raman Spectrosc. 30 (1999) 413–415.

32. S. Lu, W. Yaqian, J. Shaohua, J. Peng, C. Huang, G. Wu, L. Zhang, The contrastive studies of microwave and conventional roasting CuCl residue from zinc hydrometallurgy, in: J.-Y. Hwang, C. Bai, J.S. Carpenter, S. Ikhmayies, B. Li, S.N. Monteiro, Z. Peng, M. Zhang (Eds.), Characterization of Minerals, Metals, and Materials 2013, John Wiley & Sons Inc., Texas, 2013, pp. 529–540.

33. T. Fujimori, M. Takaoka, Direct chlorination of carbon by copper chloride in a thermal process, Environ. Sci. Technol. 43 (2009) 2241–2246.

34. T. Kosec, P. Ropret, A. Legat, Raman investigation of artificial patinas on recent bronze – part II: urban rain exposure, J. Raman Spectrosc. 43 (2012) 1587– 1595.

35. D. Persson, D. Thierry, N. Le Bozec, Corrosion product formation on Zn55Al coated steel upon exposure in a marine atmosphere, Corros. Sci. 53 (2011) 720–726.

36. [36] M.C. Hales, R.L. Frost, Synthesis and vibrational spectroscopic characterization of synthetic hydrozincite and smithsonite, Polyhedron 26 (2007) 4955–4962.

37. R.L. Frost, A. Soisnard, N. Voyer, S.J. Palmer, W.N. Martens, Thermo-Raman spectroscopy of selected layered double hydroxides of formula Cu6Al2(OH)16CO3 and Zn6Al2(OH)16CO3, J. Raman Spectrosc. 40 (2009) 645–649.

38. J. Hedberg, N. Le Bozec, I.O. Wallinder, Spatial distribution and formation of corrosion products in relation to zinc release for zinc sheet and coated preweathered zinc at an urban and a marine atmospheric condition, Mater. Corros. 64 (2013) 300–308.

39. T.H. Muster, I.S. Cole, The protective nature of passivation films on zinc: surface charge, Corros. Sci. 46 (2004) 2319–2335.

40. P. Volovitch, T.N. Vu, C. Allély, A. Abdel Aal, K. Ogle, Understanding corrosion via corrosion product characterization: II. Role of alloying elements in improving the corrosion resistance of Zn–Al–Mg coatings on steel, Corros. Sci. 53 (2011) 2437–2445.

41. I.S. Cole, N.S. Azmat, A. Kanta, M. Venkatraman, What really controls the atmospheric corrosion of zinc? Effect of marine aerosols on atmospheric corrosion of zinc, Int. Mater. Rev. 54 (2009) 117–133.

42. S. Schürz, G.H. Luckeneder, M. Fleischanderl, P. Mack, H. Gsaller, A.C. Kneissl, G. Mori, Chemistry of corrosion products on Zn–Al–Mg alloy coated steel, Corros. Sci. 52 (2010) 3271–3279.

Evaluation of Corrosion Inhibitors Effectiveness in Oilfield Production Operations

Osokogwu, U[1] and Oghenekaro .E.[2]

[1]Pursuing Ph.D degree program in Petroleum Engineering in University of Port, Harcourt,Nigeria.

[2]pursuing masters degree admission in Petroleum Eengineering in UK.

ABSTRACT

It is known that corrosion is a natural process and is impossible to prevent completely. Thus we only try to control corrosion. Even though coatings and cathodic protection are often more effective, chemical inhibitors are also widely used to reduce corrosion particularly in gas wells producing CO_2, H_2S and water. The effectiveness of the inhibitor and compatibility with produced fluids must be tested in the laboratory. Inhibitor film efficiency depends on the inhibitor concentration and contact time with the metal surface. A compact and relatively inexpensive system called High Speed Autoclave Test (HSAT) was used with corrosive gases, such as H_2S and CO_2. Using this system, the effectiveness of inhibitor was evaluated and all the variables that influences corrosion rate were easily controlled in the laboratory, in order to predict field corrosion rates. Several inhibitors were evaluated, active ingredients of those inhibitors include long chain amines, amides, and imidazoline est inhibitors were tested at the c oncentration range of 500- 10000ppm in a mixture of brine/hydrocarbon in the presence of H_2S and CO_2. In the experimental investigation, results showed that inhibitor D (imidazoline surfactant) was the most efficient (92%) at 1000ppm.

INTRODUCTION

Oil and gas production operations utilize a tremendous amount of iron and steel materials. These materials are in form of pipes, tubing, casing, pumps, valves and other accessories which are susceptible to corrosion depending on the composition and characteristics of the produced fluids. The produced fluids either in two-phase or three phase are transported through a net-work of pipelines from various sizes of tubings to central gathering stations where separation and emulsion treatment are being carried out.

However, in transportation, the internal parts of the pipelines are in constant contact with fluids and other impurities such as hydrogen-sulphide, carbondioxde and others that propagates corrosion under the operating temperature and pressure conditions.

One of the major ways of protecting the internal production pipelines in the field of operations against corrosion is by applying corrosion inhibitors. The corrosion inhibitors are evaluated in order to determine if the corrosion preventive measures applied are necessary and to know if the required life-time can be achieved with an inhibitor as effective life of corrosion inhibitors varies with the quantity of water intrusion. The purpose of the paper is to evaluate the effectiveness of commercially available corrosion inhibitors under different temperature and pressure conditions with different well effluents. A greater number of scientific studies have been devoted to corrosion inhibitors. However, most of what is known to have grown from trial and error experiment, both in the laboratories and in the field. Rules, equations, and theories to guide inhibitor development or use are very limited. By definition, a corrosion inhibitor is a chemical compound or substance that, when added in small concentration to an environment, effectively decreases the corrosion rate. The efficiency of an inhibitor can be expressed by a measure of this improvement:

$$\text{Inhibitor efficiency } (\%) = 100 \frac{(CR \text{ uninhibited} - CR \text{ inhibited})}{CR \text{ uninhibited}}$$

Where

CR uninhibited = Corrosion rate of uninhibited

CR inhibited = Corrosion rate of inhibited

In general, the efficiency of an inhibitor increases with an increase in inhibitors concentration (e.g. a typically good inhibitor would give 95% inhibition at a concentration of 0.008% and 90% at a concentration of 0.004%). Also, the reliability of assessment of the effectiveness of gas flow lines protection by inhibition depends on the method employed. Assessment based on results received by several methods make it possible to find a set of inhibitors that rate the most effective for each specific field and to develop the most optimal technology for their application. As a matter of fact, the effectiveness, or corrosion inhibitor efficiency, of a corrosion inhibitor is a function of many factors like: fluid composition, flow regime, temperature, partial pressure of CO_2 and H_2S. If the correct inhibitor and quantity is selected, then it is possible to achieve high, 90- 99% efficiency. Some mechanism of its effect are function of a passivation layer (a thin film on the surface of the material that stops access of the corrosive substance to the metal), inhibiting either the oxidation or reduction part of the redox corrosion system (anodic and cathodic inhibitors), or scavenging the dissolved oxygen.

EFFECT OF INHIBITORS

Gopal and Jepson (1995), defined inhibitor as a substance, that when added in small concentrations decrease the effect of corrosion rate. Inhibitors fall into four general categories, based on mechanism and composition, these categories are;

1.   Barrier Inhibitors
2.   Neutralizing Inhibitors
3.   Scavenging Inhibitors

Barrier Inhibitors

Barrier inhibitors form a layer on the corroding metal surface, modifying the surface to reduce the apparent corrosion rate. They represent the largest class of inhibitive substance. Absorption type inhibitors are the most common barrier layer inhibitors. In general these organic compounds are adsorbed and form a stable bond is the metal surface. The apparent corrosion rate decreases as surface adsorption is completed. Vapour phase corrosion inhibitor (VPCI) is adsorption type corrosion inhibitors with high passivation properties. These inhibitors form a stable bond with the metallic surface. Generally, they have a high vapour pressure that allows the material to migrate to distant metallic surface. Therefore, VPCI, require no direct contact with the metal surface to be protected. Conversion inhibitors also form barrier layers. They passivate the metallic surface by developing and insoluble metal oxide on the surface. Typical examples of this type of inhibitor are organic phosphates and chromates (Margarita Kharshan, 1998).

Neutralizing Inhibitors

Neutralizing inhibitors reduce the hydrogen ions in the environment. Typical neutralizing inhibitors are amines, ammonia (NH_3), and merpholine. These inhibitors are particularly effective in boiler water treatment and weak acid solutions but have been widely used on flow lines (Margarita Kharshan, 1998).

Scavenge Inhibitors

Scavenging inhibitors remove corrosive ions from solution. Well known scavenging inhibitors include hydrazine and sodium sulphite. These two inhibitors remove dissolved oxygen from treated boiler water (McMahon et al, 2005). Sodium sulphite reaction;

$$Na_2SO_3 + \tfrac{1}{2}O_2 = Na_2SO_4$$

Hydrazine reaction;

$$2\,(H_{2NN}H_2) + \tfrac{1}{2}\,O_2 = 2NH_3 + H_2O + N_2$$

Demands of Inhibitors

Pour Point

Because inhibitors are usually stored and used outdoors, they must remain liquid at low temperature. A pour point of -30°C (-20° F) is usually required. Some areas of the world may have an even lower pout point requirement (-40 to-45°C) or (-40 to-500 F). The required pour point often restates the activity and solvent systems of particular inhibitors (Hambly, 1981)

Solubility

It is dictated by the intended uses. By their way nature, inhibitors cannot be truly soluble in either hydrocarbon or water; degree of dispensability is more descriptive (Jones et al, 1996)

Performance

The end user of corrosion inhibitors will often specify a laboratory test that the inhibitor must go before a field trial or purchase will be considered. The wheel test is commonly used in the oil and gas producing industry. Therefore, many inhibitors are formulated to pass the wheel test. But in this project HSAT is used as our reference point (Hamby, 1981).

Emulsion Tendencies

The application of the inhibitors must not cause secondary problem. Batch treatment has often caused emulsions of the hydrocarbons and water that, relatively to normal operations, are externally difficult to break. In some cases, the emulations resulting from batch treatment were so severe that the surface separation equipment was literally stopped by the emulsion formed by the presence of high inhibitor concentration.

Therefore, inhibitors are specifically formulated to be non-emulsifying. Alternatively emulsion breakers (chemical) are added to formulations in small amounts to prevent emulsions (Hamby 1981).

EXPERIMENTAL PROCEDURE

Corrosion test was performed in a modified HSAT test. The HSAT tests uses an open cage spindle containing flat coupons and are rotated at different speeds, in order to generate high local shear stresses on the leading edge of the coupons. This test has been extensively used in developing corrosion inhibitors for applications where ultrahigh shear conditions caused server localized corrosion in gas pipelines.

The test has often been called a rotating cage test. In the normal procedure, a mixture of brine/hydrocarbon is added to the autoclave. After purging/ evaluation to remove oxygen, inhibitor is added at a specific concentration. The stirrer in then turned in and the pressure and temperature are adjusted to test conditions. At completion of the test, the apparatus are allowed to cool. Coupons are then removed, inspected, and re-weighed. A corrosion rate in calculated for a specific test time and weight loss. In this work, corrosion test are reported for experiment at 188^0C (370^0 F), 10.3Mpa (500psi) and 232°C (450° F), 2.88Mpa (420psi), and open cage spindle containing flat coupons are used.

Each coupon is used in the spindle. Each coupon has a surface area of 11.3cm². The ratio of volume of liquids surface of the coupons is 133cm. At 200ppm, the linear velocity of the coupon in the cage is 6.65m/sec (21.82ft/sec). Brine was added to the autoclaves, after purging/evacuation to remove O_2, inhibitor was added at specific concentration, the stirrer was turned on, and the pressure was adjusted to 2.884mpa (420psi). The partial pressure of CO_2 in the test was 2.88 mpa (420pisa). The concentration of the H_2S used in the test was 12ppm. After reaching the test temperature of 232°C (4500 F), the test was contained for 18hrs. After this time, the system was allowed to cool and the coupons were removed, inspected and re-weighted after cleaning.

Corrosion rate was calculated for the specific test concentrations. Inhibitors performance was assessed on the basis of corrosion rate relative to the blank-corrosion efficiency. The brine composition used in our test at 1880C (3700 F), 10.3mpa (1500psi) is shown in Table 2.

An 80/20 mixture of brine/condensate was used as the test fluid. Testing was conducted using 10.3Mpa (1500psi) of CO_2, 0.021 Mpa (3psi) H_2S, and temperature of 188°C (3700 F). Testing was performed at both 2000 and 500rpm to stimulate high flow rate and low flow rate condition, respectively. Testing was done on the high speed autoclave test (HSAT) described earlier. The linear velocity at 500rpm is 1.66m/sec (5.45ft/sec). The inhibitors used in the experiment were oil soluble corrosion inhibitors. Corrosion inhibitors A was a cyclic amine based corrosion inhibitor. Corrosion inhibitors B was amido-imidazoline based corrosion inhibitors. Corrosion inhibitors C was amine imidazoline based corrosion inhibitor. Corrosion inhibitors D was an imidazoline surfactant based corrosion inhibitors.

Discussion and Analysis of Results

The results of test using different concentration in high speed autoclave test at 2000ppm, 232°c (4500 F), 2.88mpa (420psi) CO_2, and 12ppm H_2S are shown in figures 1. The corrosion rate without corrosion inhibitors is 4.01mm/yr. Over 90% protection is obtained using 1000ppm of corrosion inhibitor A. The amount of corrosion inhibitors needed to achieve this amount of protection is large. This is characteristic of high temperature gas flow line situation where blank corrosion rate are lower due to the formation of an Iron carbonate scale but the concentrations of inhibitors needed to treat for corrosion are higher. The results of experiment on corrosion inhibitors performance with high speed autoclave testing under a pressure of 10.3mpa (1500psi) CO_2, 0.021mpa (3psi) H_2S with a rotating cage of 2000rpm in an 80/20 mixture of brine and condensate at 1880 c (3700 F) are shown in figures 2.

The test duration is 18hrs. The uninhibited rate in the test is 5.08mm/yr (200mpy). Corrosion inhibitors A, an oil soluble water dispersible corrosion inhibitors that has performed well in laboratory test at 2320 c (4500 F), 2.88mpa (420psi) CO_2, and 12p, H_2S did not do well in this test. Corrosion inhibitor B also did not do well. At high concentration using 1000ppm, corrosion inhibitors C provided above 87% protection. A new formulation corrosion inhibitors D, gave the best protection under these severe conditions. Corrosion inhibitor of above 92% using a dosage of 1000ppm was obtained in our laboratory test. Longer duration test may give lower corrosion rates, due to resistance developed by the gas. In order to mimic low shear conditions, high speed autoclave test at 500rpm were conducted at carbon dioxide partial pressures of 10.3mpa (1000psi), 0.02mpa (3psi) hydrogen sulfide in a 80/20 mixture of brine and condensate at 180°C (370° F).

The results of test with corrosion inhibitor C and corrosion inhibitors D are shown in figure 3. The uninhibited rate is higher (13.77mm/yr.) then in systems under higher shear (5,08mm/yr). This is sometime seen in corrosion tests involving hydrogen sulphide. A 90% corrosion inhibitor protection is achieved using corrosion inhibitors C at a dosage of 1000ppm. Higher corrosion inhibitors efficiency of above 96% can be obtained using the new corrosion inhibitors D at a dosage of 500ppm.

Table 1: Brine used in testing at 232°C (450°F) and a partial pressure 2.884Mpa (420Psi) carbon dioxide

Component	Concentration (mg/L)
Nacl	5100
$CaCl_2.2H_2O$	600
$MgCl_2. 6H_2O$	70

Table 2: Composition of Brine used in testing at (370°F) 10.3Mpa (1500Psi)

Component	Concentration (mg/L)
Nacl	144810
$CaCl_2.2H_2O$	18510

Experiment1: Results

Inhibitor A

Concentration	Corrosion Rate	Remarks
0	4	
500	1.3	
1000	0.4	
2000	0.2	

Experiment 2 at 188° C: Results

Inhibitor A

Concentration	Corrosion Rate	Remarks
0	10	
500	7	
1000	0.4	
2000	0.2	

Inhibitor B

Concentration	Corrosion Rate	Remarks
0	4.5	
500	3.8	
1000	3.3	
2000	3.25	

Inhibitor C

Concentration	Corrosion Rate	Remarks
0	4.5	
500	3.28	
1000	3.10	
2000	2.8	
5000	0.8	
10,000	0.5	

Inhibitor D

Concentration	Corrosion Rate	Remarks
0	4.5	
500	0.23	
1000	0.21	
2000	0.20	
5000	0.19	
10,000	0.88	

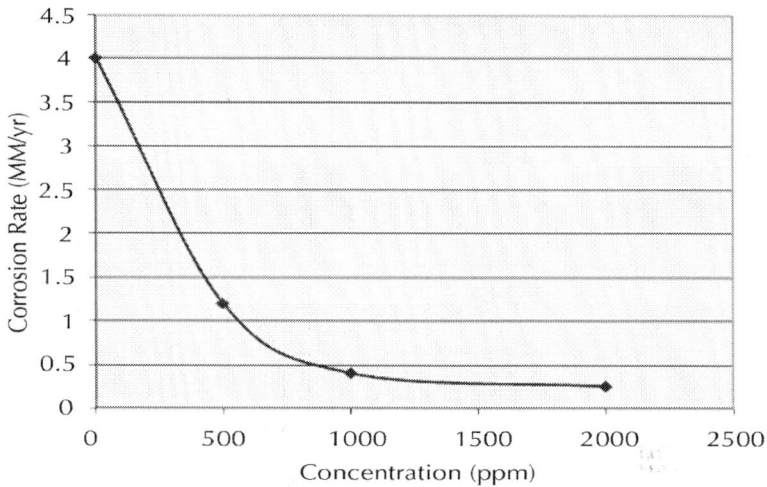

Figure 1: A plot of corrosion rate vs Concentration (ppm); Results of Corrosion Test at 232°C (450° F), 2000rpm with a partial pressure of 2.884Mpa (420Psia) CO2 and 12ppm concentration of H_2S.

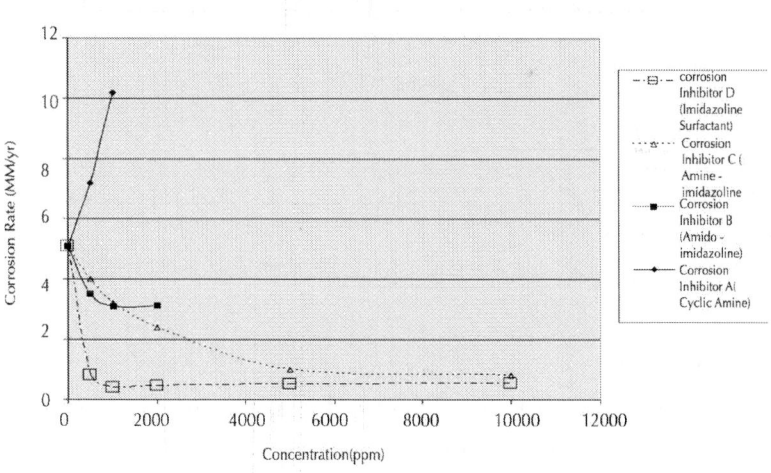

Figure 2: General Corrosion Rate from High Speed Autoclave Test at 188°C (370°F) and 10.3Mpa (1500psia) CO_2, 3psia H_2S in a 80/20 brine/oil mixture at 2000pm.

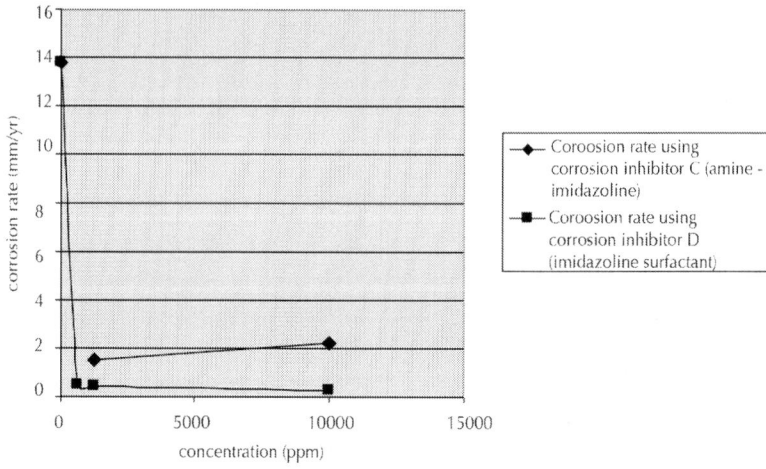

Figure 3: General Corrosion Rates from High Speed Autoclave Test at 188°C (370°F) and 10.3Mpa (1500psia) CO_2, 0.021Mpa (3psia) H_2S in a 80/20 brine/oil mixture at 500rpm.

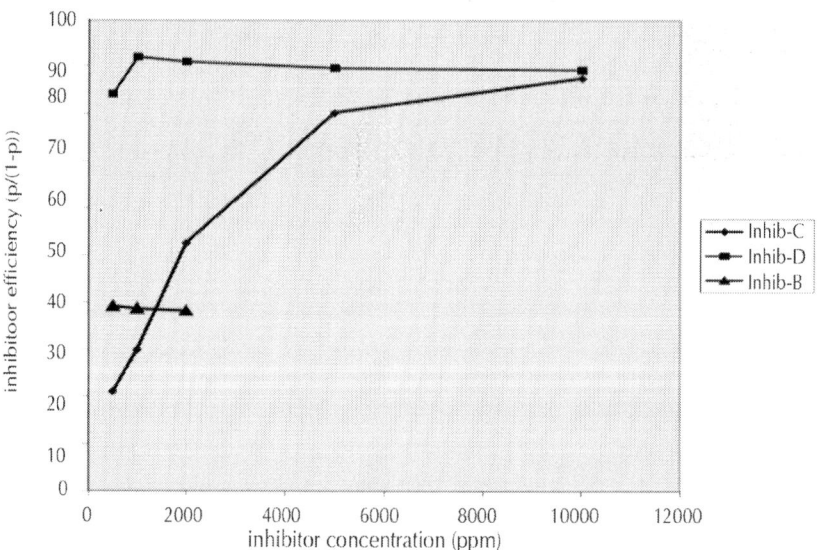

Figure 4: Graph showing inhibitor effectiveness vs Inhibitor concentration.

APPENDIX

Table 3: Molecular composition of a typical Nigeria Reservoir fluid

Components	
Nitrogen	0.21
Carbon dioxide	2.34
Methane	40.66
Ethane	2.16
Propane	0.14
Iso Butane 0.57	0.05
Normal Butane	
Iso Pentane	0.33
Normal Pentane	0.01
Hexanes	0.98
Heptanes +	52.55
Total	100.00%
Molecular Weight	149.03
Heptanes +Molecular Weight	265.00

Corrosion Rate Calculation

The corrosion rate is calculated by the following formula:

$$\text{Corrosion rate (mpy)} = \frac{\text{Coupon weight loss (g)} \times 2.23 \times 10^4}{\text{Total exposed area of coupon (in}^2) \text{ Exposure time (days)} \times \text{metal density}}$$

Corrosion Rate in mpy	
Low	< 1.0
Moderate	1.0 – 4.9
High	5.0 – 10
Severe	> 10

Pitting Rate Calculation

The pitting rate can be calculated by the following formula.

$$\text{Pitting rate in mils per year (mpy)} = \frac{\text{Pitting depth (mils) x 365}}{\text{Exposure time (days)}}$$

Mole (%)

In addition, this crude is waxy.

Table 4: Typical Analysis of condensate water

Content	A Field from Niger Delta (ppm)	Niger Delta Field (ppm)
Sodium Chloride	304	254
Sodium Bicarbonate	138	161
Calcium Sulphate	26	30
Calcium Chloride	19	7
Magnesium Chloride	20	21
Iron	220	255
Organic acids as acetic acid	480	450

REFERENCES

1.   Dunlop, A.K., H. L. Hassel, P. R. Rhodes, " Fundamental considerations in sweet gas well corrosion", Paper No. 46, Corrosion/83, NACE International, 1983

2.   Mc mahem, A.J., Martin,J.W and L. Harris, "Effects of sand and interfacial adsorption loss on corrosion inhibitor efficiency", Corrosion/2005, Paper No. 05274, 2005

3.   De waard, C. U. Lotz and D. E. Millian, "Carbonic acid corrosion of steel", Corrosion 31(5), 1975

4.   Ernest W. Klechka, "How to predict and control sweet oil well corrosion", Oil and Gas Jornal, Vol. 50, P.116 – 118,151., 2001

5.   Greenwell J. T, " Effect of CO2 on corrosion of pipelines"1952

6.   G. Schmitt, C. Bosch, U. Pankoke, W. Bruckhoff, G. Siegmund, " Evaluation of critical flow intensities for FILC in sour gas production", Paper No. 46, Corrosion/98, 1998

7.   G. Schmitt, W. Bruckhoff, K. Fassler, G. Blummel, " Flow loop vs rotating probies – correlation between eexperimental results and services application", NACE CORROSION, Paper No. 23, 1990

8.   Gopal and Jepson W. P,, " The flow characteristics in horizontal slug flow", Parer presentation at 3rd international conference on multi – phase flow, 1995

9.   Horsup David , " Corrosion inhibition, corrosion chemistry", ACS symposium series, Vol. 89, P. 316, 2007

10. J. A. Dougherty, S. Ramachandran, B, Short, " Does shear stress have an effect on corrosion in sour gas production", NACE CORROSION/2001, Paper. No. 1069, 2001

11. Jones, G. L. Farrar, Sydberger , " Combatting corrosion in oil and gas wells", Oil and Gas Journal, Vol. 50, P. 106 – 109,111,113, 1996

12. King T, C. U. Lotz and D. E. Millians, " The methodology of corrosion inhibitor development for CO_2 systems", CORROSION, Vol. 45, Paper No. 10, 1991

13. K. Kennedy, John L. (1993)," Oil and gas pipeline fundamentals", 2nd edition, Pennwell books, P.366, 1993

14. Lotz C. U, and T. Sydberger , " CO2 corrosion in carbon steel and 13 Cr Steel in particle Laden fluid", CORROSION, 44(11), 1990

15. Margarita Kharshan, Alla Furman, " Incorporating vapour corrosion inhibitors (VCI) in oil/gas pipelines", NACE, Paper No. 236, 2000

16. Peabody A. W, " Control of ppipeline corrosion", Houston T X, NACE, P.190, 1978

17. Sydberger T, " Flow dependent corrosion mechanism, damage charateristics, and control", British Corrosion Journal,P. 83-89, 1987

18. S. Papavinasam, R. H. Hausler, C. I. Cruz, H. Sutano, "laboratory studies on flow induced localized corrosion in CO_2/H_2S environment, and development of test methodology", NACE INTERNATIONAL, Paper No. 1, 1999

19. T. W. Hamby, " Development of high pressure sour gas technology", CORROSION, Parper No. 8309, P. 119, 1981

20. Videm K. and A. Dugstad, " Corrosion of carbon steel in an aqueous CO_2 enviroment: part 1", Material performance, P. 63, 1989

Sweet Corrosion Inhibition on API 5L-B Pipeline Steel

Mahmoud Abbas Ibraheem,[1] Abd El Aziz El Sayed Fouda,[2]
Mohamed Talaat Rashad,[3] and Fawzy Nagy Sabbahy[1]

[1]Department of Metallurgical and Material Engineering, Faculty of
Petroleum and Mining Engineering, Suez Canal University, Suez, Egypt

[2]Department of Physical Chemistry, Faculty of Science, Mansoura
University, Mansoura, Egypt

[3]Department of Integrity Management, United Gas Derivatives
Company (UGDC), Port Said, Egypt

ABSTRACT

Corrosion inhibition and adsorption behavior of two triazole derivatives
on API 5L-B carbon steel in CO_2-saturated 3.5% NaCl solutions

was investigated using potentiodynamic polarization, EIS, and EFM techniques. Specimen surfaces were characterized using SEM, EDX, and XRD. Results show that the two compounds are mixed-type inhibitors and inhibition efficiency increases with increasing concentrations. Adsorption of the two compounds chemisorption and obeys Langmuir adsorption isotherm. Activation energy and thermodynamic parameters were calculated. Surface analyses confirm the formation of iron nitrides on the metal surface which supports results obtained from previous techniques.

INTRODUCTION

Carbon dioxide which is present naturally in oil and gas wells is injected purposely into wells to enhance oil recovery. CO_2 corrosion, also known as "sweet corrosion," is one of the major problems in oil and gas industry, costing billions of dollars every year. Great efforts must be expended in corrosion control for safety, business, and environmental considerations.

Sweet corrosion is caused by the presence of carbon dioxide (CO_2) dissolved in water to form carbonic acid (H_2CO_3). Corrosion increase as the concentration of CO_2, system pressure, and temperature increase. This corrosion is typically slow, localized and results in pitting attack. Pits are very difficult to detect because of their tiny size and the corrosion products that cover them.

In oil and gas production and processing industries, corrosion inhibitors have always been considered the first line of defense against internal corrosion. Inorganic inhibitors, such as sodium arsenite (Na_2HAsO_3) and sodium ferrocyanide, were used in early days to inhibit carbon dioxide (CO_2) corrosion in oil wells, but the treatment frequency and effectiveness were not satisfactory. This led to the development of many organic chemical formulations that frequently incorporated film-forming amines and their salts.

In this work, the corrosion inhibition and adsorption behavior of Itraconazole and Fluconazole compounds on API 5L-B carbon steel in CO_2-saturated 3.5% NaCl solutions was investigated using potentiodynamic polarization, electrochemical impedance spectroscopy, and electrochemical frequency modulation techniques.

The surfaces of the samples were characterized using SEM, EDX, and XRD techniques. The effect of temperature on the corrosion rates and inhibition process was also investigated.

EXPERIMENTAL

Materials Preparation

Experiments were conducted using a conventional three-electrode 200 mL cell assembly with the counter electrode made of platinum and saturated calomel electrode (SCE) reference electrode. API 5L-B carbon steel cut from its parent pipe was used as the test material and has the chemical composition shown in Table 1.

Table 1: Chemical composition of the API 5L-B carbon steel

Element	C	Mn	P_{max}	S_{max}	Fe
Composition (Wt%)	0.28	1.2	0.03	0.03	Balance

The steel was cut into coupons of dimension 1 × 1 × 0.5 cm. the coupons were degreased with acetone, air dried and embedded into two-component epoxy resin and mounted in a glass holder. A copper wire was soldered to the rear side of the coupon as an electrical connection. The exposed surface of the electrode (of area 1 cm²) was wet polished with silicon carbide abrasive papers up to 1200 grit, rinsed with ethanol, and air dried [1, 2]. This was used as the working electrode during the electrochemical test.

The test medium was 150 mL 3.5% NaCl solutions saturated with carbon dioxide gas at atmospheric pressure. Before the test, the system was deaerated by flushing with CO_2 gas for two hours and kept saturated with CO_2 by a continuous purging of the gas [3]. The gas exit was sealed with distilled water. The temperature was maintained within ±1°C in all experiments by placing the cell into a thermostatic water path. The pH was monitored with PB-10 Sartorius pH meter that was carefully calibrated with two buffer solution (pH 4 and pH 7). The pH was in all experiments.

The two inhibitors, Itraconazole and Fluconazole, are triazole compounds, have the structure shown in Figure 1. Itraconazole and Fluconazole are Antifungal drugs obtained from APEX and SEDICO as trade names, Itrapex and Flucoral, respectively. Concentrations of 0.1, 0.5, 1, 5, 10, 15, 20, and 25 ppm were used.

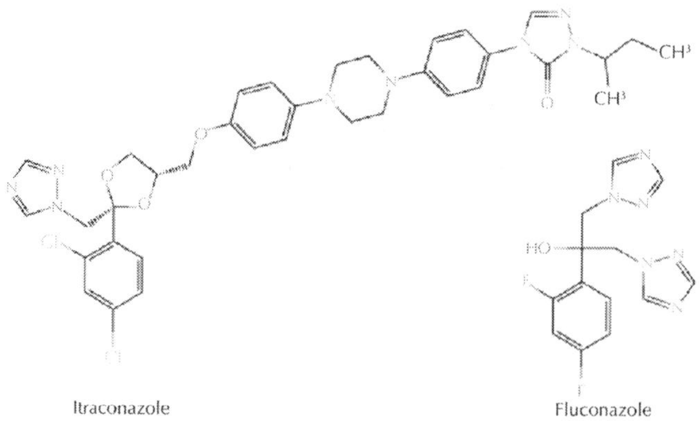

Itraconazole   Fluconazole

Figure 1: Molecular structures of the tested compounds.

Potentiodynamic polarization experiments were carried out using VoltaLab-PGZ100 connected to personal computer utilizing VoltaMaster-4 software, version 7.08, while EIS and EFM were carried out using Gamry instrument G750-Series Potentiostat / Galvanostat / ZRA with a Gamry framework system based on ESA400. Surface characteristics were investigated using Jeol JSM-5410 for SEM, Oxford In CA PentaFETX3 for EDX, and PANalytical X Pert Pro NC 4022 for XRD.

Experimental Procedure

The free corrosion potential (versus SCE) was followed after immersing the working electrode in the test solution until the potential stabilized within ±1 mV. This was followed by EIS test, performed over the frequency range of 100 kHz–100 mHz with a signal amplitude perturbation of 5 mV.

The charge transfer resistance (R_{ct}) was obtained from the diameter of the semicircle in the Nyquist plot, and the inhibition efficiency determined from the relationship:

$$IE\% = \frac{R_{ct} - R_{ct}^{\circ}}{R_{ct}} \times 100, \tag{1}$$

where R_{ct}° and R_{ct} are the uninhibited and inhibited charge transfer resistance, respectively. The characteristic frequencies f_{max} were obtained from the semicircles maxima and used to calculate the capacitance from the equation:

$$C_{dl} = \frac{1}{2\pi f_{max}} \frac{1}{R_{ct}} \tag{2}$$

The potentiodynamic sweeps were conducted at a constant pH and sweep rate of $5\,mVs^{-1}$, and the corrosion current density, i_{corr}, determined graphically by extrapolating the linear Tafel segment of the cathodic curves to the E_{corr} (versus SCE). From the measured i_{corr} values, the inhibition efficiency was calculated using the relationship:

$$IE\% = \frac{i_{corr}^{\circ} - i_{corr}}{i_{corr}^{\circ}} \times 100 \tag{3}$$

where i_{corr}° and i_{corr} are the uninhibited and inhibited corrosion current densities, respectively. The solution and metal coupon were changed after each sweep.

RESULTS AND DISCUSSION

Potentiodynamic Polarization Measurements

Figures 2(a) and 2(b) show the polarization curves for API 5L-B in CO_2-saturated 3.5% NaCl solutions at 25°C and pH 3.8, in the absence and presence of Itraconazole and Fluconazole, respectively. Polarization parameters obtained from the curves are as shown in Table 2. The introduction of the inhibitors shifted E_{corr} (versus SCE) towards the negative region, and reduced i_{corr}. These changes are dependent on the inhibitor concentration reaching the maximum change at 15 and 20 ppm for Itraconazole and Fluconazole, respectively.

Table 2: Polarization data obtained in the absence and presence of inhibitors at 25°C

Comp.	Conc., ppm	$-E_{corr}$ versus SCE, mV	i_{corr}, $\mu A \cdot cm^{-2}$	$-_{c'}$ mV·dec^{-1}	$_{a'}$ mV· dec^{-1}	θ	IE, %	C.R, $\mu m \cdot y^{-1}$
Blank	0	623	289	367	110			3387
Itraconazole	0.1	631	230	276	100	0.204	20.4	2691
	0.5	621	152	277	90	0.474	47.4	1784
	1	671	100	230	92	0.654	65.4	1179
	5	670	90	255	91	0.689	68.9	1062
	10	696	62	274	98	0.785	78.5	732
	15	683	24	137	85	0.917	91.7	282
Fluconazole	0.1	615	221	330	110	0.235	23.5	2592
	0.5	667	184	342	128	0.363	36.3	2159
	1	628	144	147	80	0.502	50.2	1691
	5	670	81	181	96	0.720	72.0	958
	10	684	63	157	82	0.782	78.2	744
	15	674	44	136	77	0.848	84.8	519
	20	680	28	122	67	0.903	90.3	336

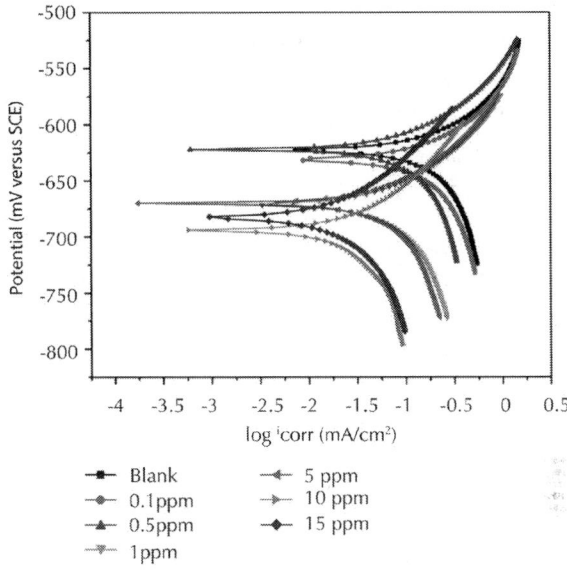

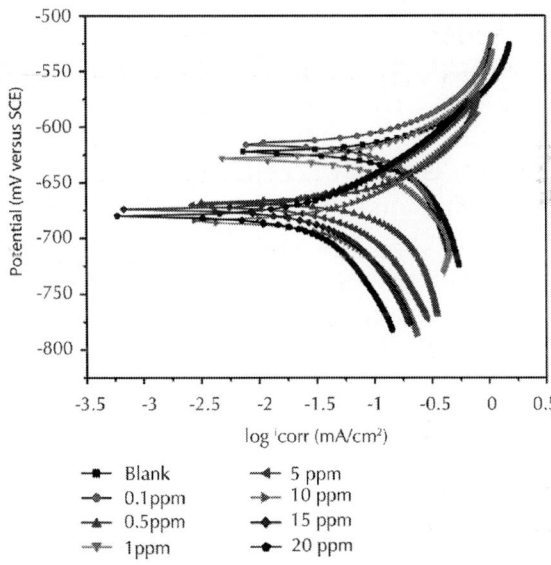

Figure 2: (a) Polarization curves in the absence and presence of Itraconazole at 25°C. (b) Polarization curves in the absence and presence of Fluconazole at 25°C.

Li, John, and others [4–6] show that if the displacement in the corrosion potential is more than ±85 mV with respect to the blank, then the inhibitor is labeled as cathodic or anodic type; otherwise inhibitor is treated as a mixed type. In our study, the maximum displacement was 73 and 62 mV for Itraconazole and Fluconazole, respectively. This indicates that the two inhibitors are mixed types. This means the inhibitors reduce the corrosion current without causing considerable change in the corrosion mechanism. This means also that the addition of these inhibitors to the solution reduces the anodic dissolution and retards the cathodic hydrogen evolution reaction [4].

However, the shift in the corrosion potential indicates that the inhibition for this system is probably due to the active sites blocking effect. This could be explained as the inhibitor molecules adsorb on the metal surface, form a protective film, and block the available reaction sites [6]. The formation of inhibitor film on the metal surface reduces the active surface area exposed to the corrosive medium and delays the hydrogen evolution and metal dissolution. This provides considerable protection to metal surface in the solution. The surface coverage of the inhibitor molecules on the metal surface increases with increasing the inhibitor concentration. Potentiodynamic polarization curves in Figures 2(a) and 2(b) exhibit no steep slope in the anodic range, meaning that no passive films are formed on the metal surface [7]. As per the corrosion theory the shift in the cathodic curves reveals that the corrosion process is mainly accelerated by the cathodic reactions. Accordingly, these inhibitors seem to retard hydrogen evolution reaction via blocking the active reaction sites. The inhibition efficiency of the two inhibitors found increases with increasing inhibitor concentration and reaches its highest value at 15 and 20 ppm for Itraconazole and Fluconazole, respectively. The increase in the inhibition efficiency with the increase in inhibitor concentration is attributed to the increased surface coverage by the inhibitor molecules as the concentration is increased [5].

Electrochemical Impedance Spectroscopy (EIS) Measurements

The Nyquist plots obtained from the EIS measurements for API 5L-B in CO_2-saturated 3.5% NaCl solutions at 25°C and pH 3.8, in the absence and presence of Itraconazole and Fluconazole are shown in Figures 3(a) and 3(b), respectively. It is appeared from the figures that the shapes of Nyquist plots for the two inhibitors are similar. At each concentration, the shape of the Nyquist plot remained the same with one depressed semicircle in all experiments, indicating that almost no change in the corrosion mechanism occurs due to the inhibitor addition. The shape of the Nyquist plot looks comparable to that obtained by John, Joy, Prajila and others [7–9].

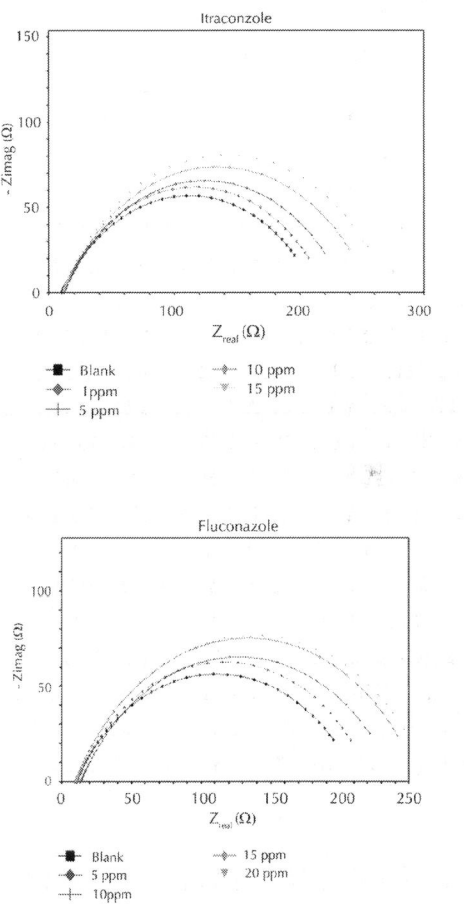

Figure 3: (a) Nyquist plot in the absence and presence of Itraconazole at 25°C. (b) Nyquist plot in the absence and presence of Fluconazole at 25°C.

These diagrams indicate that the impedance spectra consist of one capacitive loop at high frequency. The introduction of the inhibitors was observed to increase the capacitive semicircles and this increase is dependent on the concentration of the inhibitors. This was ascribed to the double layer capacitance and charge transfer resistance and indicates inhibition of the corrosion process [10].

Various parameters obtained from impedance measurements as R_{ct}, C_{dl}, and i_{corr} are given in Table 3. On the introduction of the inhibitors, R_{ct} values increase and the calculated C_{dl} values decrease, as shown in Table 3. These changes increase with increasing inhibitor concentration.

Table 3: Parameters obtained from EIS in the absence and presence of inhibitors at 25°C

Comp.	Conc., M	R_s, $\Omega\,cm^2$	$Y \times 10^{-4}$, $\mu\Omega^{-1}\,sn$ cm^{-2}	n	R_{ct}, $\Omega\,cm^2$	$C_{dl} \times 10^{-4}$, $\mu F cm^{-2}$	θ	IE, %
Blank	0	9.81	9.93	0.644	203.2	4.10		
Itraconazole	1	10.85	8.20	0.674	210.1	3.50	0.033	3.3
	5	11.92	8.21	0.669	224.9	3.56	0.096	9.6
	10	11.38	7.49	0.684	245.7	3.43	0.173	17.3
	15	10.80	6.86	0.698	260.8	3.26	0.221	22.1
Fluconazole	5	10.89	8.64	0.676	212.7	3.84	0.045	4.5
	10	13.49	8.95	0.661	227.6	3.96	0.107	10.7
	15	9.56	6.90	0.689	249.0	3.12	0.184	18.4
	20	10.17	7.59	0.682	256.5	3.54	0.208	20.8

Increasing of R_{ct} values could be attributed to the formation of a protective electrochemical double layer on the metal-solution interface [10, 11]. Where C_{dl} values show opposite trend; that is, they decrease with increasing inhibitor concentrations; this could be due to an increase in thickness of the double layer [12]. This double layer was formed by the adsorption of the inhibitor molecules at the metal/solution interface and replaced water molecules gradually. These observations support the idea that the corrosion of API 5L-B is controlled by a charge transfer process [9, 13].

The inhibition efficiency, calculated from EIS results, shows the same trend as those obtained from polarization measurements. The difference in inhibition efficiency of the two methods may be due to the different surface status of the electrode in two measurements. EIS were performed at the rest potential, while in polarization measurements the electrode potential was polarized to high over potential, non-uniform current distributions, resulted from cell geometry, solution conductivity, counter and reference electrode placement, and so forth, will lead to the difference between the electrode area actually undergoing polarization and the total area [14].

Electrochemical Frequency Modulation (EFM) Measurements

EFM experiment was carried out assuming that the reaction is under either diffusion control or activation control and not using the passivation model since API 5L-B does not form passive films at these conditions. Results obtained from EFM measurements are intermodulation spectra shown in Figures 4, 5, and 6 for the blank, Itraconazole, and Fluconazole, respectively, and data obtained are shown in Table 4. The results show that the introduction of the inhibitors reduced i_{corr} and this effect increase with increasing concentrations. The inhibition efficiency, IE (%), calculated from (4). The causality factors in Table 4 was close to the theoretical values according to the EFM theory [15], should guarantee the validity of Tafel slopes and corrosion current densities.

Table 4: EFM data obtained in the absence and presence of inhibitors at 25°C

Comp.	Conc., M	i_{corr}, μA·cm⁻²	a', mV·dec⁻¹	c', mV·dec⁻¹	CF-2	CF-3	θ	IE, %	C.R, μmy⁻¹
Blank	0	230	99	183	1.9	3.5	0.000	0.0	2672
Itraconazole	1	186	93	179	1.9	3.2	0.191	19.1	2164
	5	166	91	175	1.9	3.2	0.278	27.8	1930
	10	153	89	166	1.9	3.6	0.335	33.5	1778
	15	143	81	149	1.9	3.8	0.378	37.8	1668
Fluconazole	5	213	101	185	1.9	3.3	0.074	7.4	2479
	10	191	94	190	1.9	2.8	0.170	17.0	2222
	15	182	94	180	1.9	3.4	0.209	20.9	2115
	20	174	90	173	1.9	3.1	0.243	24.3	2026

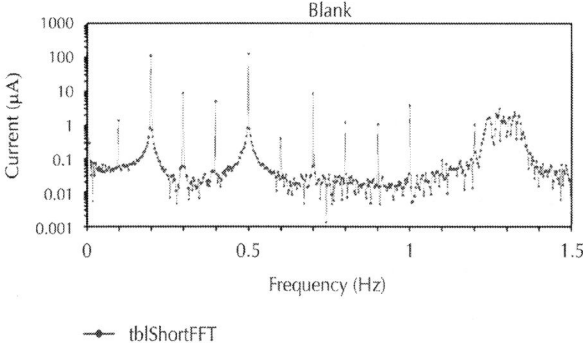

Figure 4: EFM spectra for the blank "in the absence inhibitors" at 25°C.

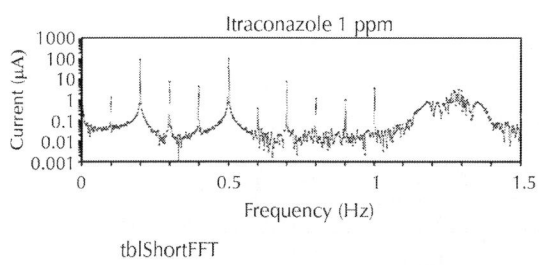

(a)

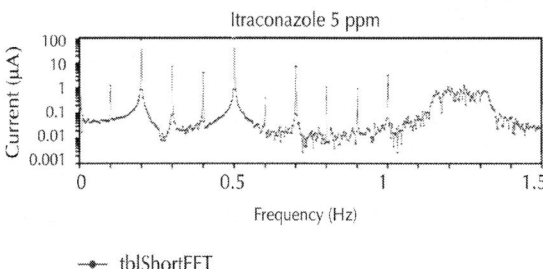

(b)

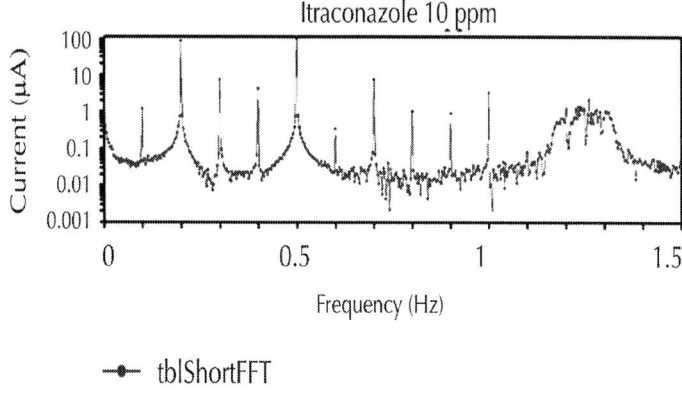

(c)

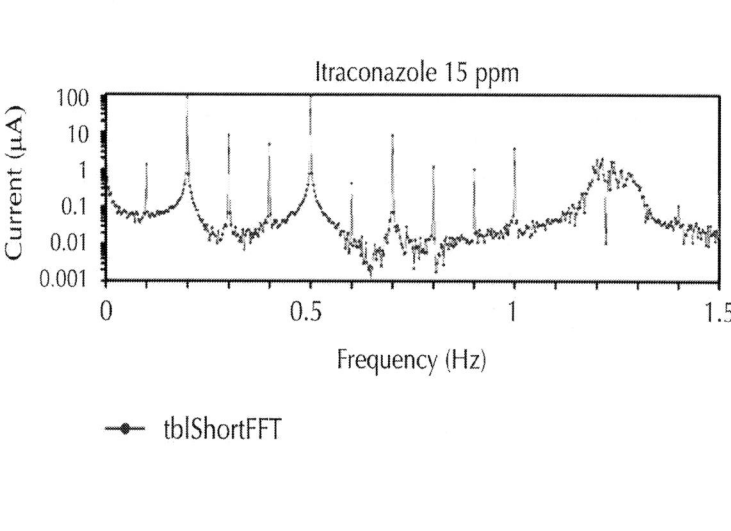

(d)

Figure 5: EFM Spectra in the presence of different concentrations of Itracon-azole at 25°C.

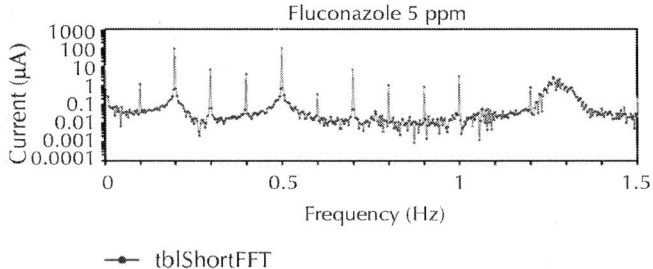

(a)

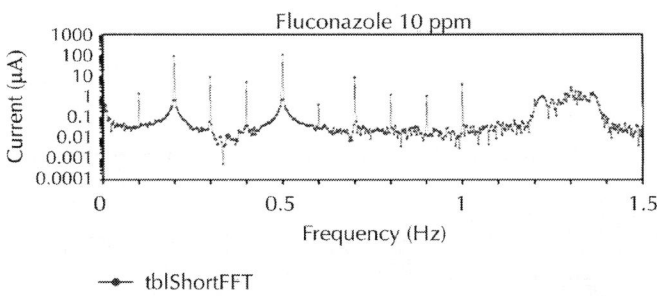

(b)

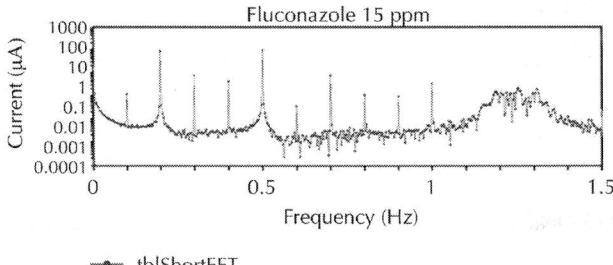

(c)

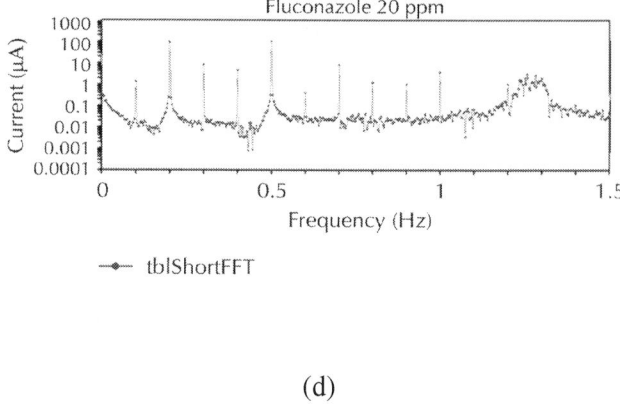

(d)

Figure 6: EFM Spectra in the presence of different concentrations of Fluconazole at 25°C.

Adsorption

Several adsorption isotherms were assessed and the Langmuir isotherm was found the most suitable one fit for the present case, Figure 7. The following equation was applied:

$$\frac{C}{\theta} = \frac{1}{K_{ads}} + C \tag{4}$$

where θ is the surface coverage ($\theta = $ IE (%)/100), C the inhibitor concentration, K the adsorption coefficient, which represents the adsorption-desorption equilibrium constant. The obtained and K_{ads} values are 9.12×10^5 and 3.90×10^5 for Itraconazole and Fluconazole, respectively. These high values reflect the high adsorption ability of the two inhibitors on the metal surface [16–18]. The standard free energy of adsorption, ΔG°_{ads}, can be calculated from K_{ads} via the following equation:

$$K_{ads} = \frac{1}{55.5} \exp\left(\frac{-\Delta G^{\circ}_{ads}}{RT}\right). \tag{5}$$

The value 55.5 in the above equation is the molar concentration of water in solution in mol/L. Thermodynamic parameters for the adsorption are given in Table 5. The large negative values of ΔG°_{ads} in the table suggested that the adsorption process takes place spontaneously [19–21] and the adsorbed layer of the inhibitor on the metal surface is highly stable [22].

Table 5: Thermodynamic parameters for adsorption of inhibitors at 25°C

Comp.	Conc., ppm	Conc. $\times$ 10^6, M	$C/\theta \times 10^6$	$K \times 10^5$	$-\Delta G^{\circ}$, kJ·Mol^{-1}
Itraconazole	0.1	0.142	0.696		
	0.5	0.711	1.5		
	1	1.42	2.17	9.12	44.0
	5	7.11	10.3		
	10	14.2	18.1		
	15	21.3	23.2		
	0.1	0.327	1.39		
	0.5	1.63	4.49		
Fluconazole	1	3.27	6.51	3.90	41.9
	5	16.3	22.7		
	10	32.7	41.8		
	15	49.0	57.8		

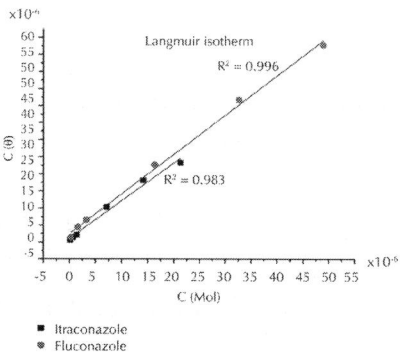

Figure 7: Langmuir adsorption isotherm for Itraconazole and Fluconazole inhibitors at 25°C.

It is well known that values of ΔG°_{ads} of the order of $20\,kJmol^{-1}$ or lower indicate that the adsorption process is physisorption, that is, electrostatic interaction between charged molecules and charged metal, while those of the order of $40\,kJmol^{-1}$ or higher indicate that it is chemisorption, that is, charge sharing or a transfer from the inhibitor molecules to the metal surface to form a coordinate type bond (covalent bond) [23–27]. Accordingly, the values of ΔG°_{ads} obtained in the present study indicate that the two inhibitors adsorbed on the metal surface by chemisorption. The same result was obtained by Tang et al. [16]. The effectiveness of Itraconazole and Fluconazole as inhibitors can be ascribed to the adsorption of their molecules on the metal surface. This adsorption could be attributed to the presence of nitrogen atoms in their molecular structures [26, 28, 29]. These atoms form iron-nitrogen coordination bonds, by a p-electron interaction between the p-electron in the head group and iron. Adsorption may also be due to the coulombic attraction (physical adsorption) [3].

Effect of Temperature

The effect of temperature on the corrosion process in the absence and presence of Itraconazole and Fluconazole is represented in Figure 8. There is a clear acceleration of both the anodic and cathodic reactions with an increase in temperature. Similar curves have been reported previously [30, 31]. The polarization parameters are listed in Table 6. It can be seen that i_{corr} increased with temperature in the uninhibited and inhibited solutions. This is due to the acceleration of all the processes involved in corrosion: electrochemical, chemical, transport, and so forth with increase in temperature.

Table 6: Effect of temperature in the absence and presence of inhibitors

Comp.	Temp., °C	$-E_{corr}$ versus SCE, mV	i_{corr} μA cm^{-2}	$-\beta_c$, mV·dec^{-1}	$_a$, mV·dec^{-1}	θ	IE, %	C.R., μm·y^{-1}
Blank	25	622.7	289	367	110			3387
	35	663	313	335	118			3663
	45	700	323	230	101			3785
	55	694	340	298	125			3978
Itraconazole	25	683	24	137	85	0.917	91.7	282
	35	666	94	132	77	0.700	70.0	1106
	45	689	161	277	112	0.502	50.2	1884
	55	717	213	355	132	0.374	37.4	2498
Fluconazole	25	674	44	136	77	0.848	84.8	519
	35	685	145	205	95	0.537	53.7	1697
	45	708	216	276	114	0.331	33.1	2528
	55	690	301	169	88	0.115	11.5	3520

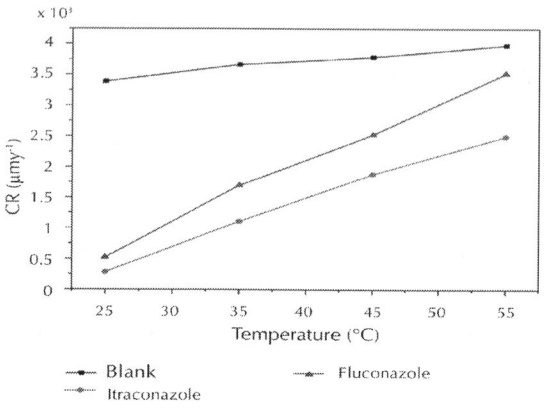

Figure 8: Effect of temperature on the corrosion rate in the absence and presence of inhibitors.

The apparent activation energies $\left(E_a^*\right)$ for the corrosion process in the absence and presence of inhibitors were calculated from the Arrhenius equation as follows:

$$K = A\exp-\frac{E_a^*}{RT}$$ (6)

where K is the corrosion rate, A is a constant, R is the universal gas constant ($R=8.314$ J·mol^{-1}K^{-1}), and T is the absolute temperature.

Figure 9 presents the Arrhenius plots and E_a^* values in the absence and presence of inhibitors as calculated from the above equation are listed in Table 7. E_a^* values are 4.40 kJ.mol^{-1} in the absence of the inhibitors, and 61.75 kJ·mol^{-1} and 53.84 kJ.mol^{-1} in the presence of Itraconazole and Fluconazole, respectively. The presence of inhibitors increases the activation energies of the metal dissolution by adsorption of their molecules on the metal surface, forming a double layer which is considered a barrier [18, 28, 32, 33]. The higher E_a^* can be correlated with the increased thickness of the formed double layer, which increases the activation energy of the corrosion process [4, 16, 23,

34]. The E_a^* values in the presence of Itraconazole and Fluconazole are nearly similar, suggesting that the two inhibitors inhibit the corrosion process with similar mechanisms. It is also indicated that the whole corrosion process is controlled by surface reaction, since the activation energy (E_a^* of the corrosion process is over 20 kJmol⁻¹ [35].

Table 7: Activation parameters in the absence and presence of inhibitors

Comp.	Conc., ppm	Temp., °C	i_{corr}, μA·cm⁻²	ΔH^*, kJ·mol⁻¹	ΔS^*, J·mol⁻¹ K⁻¹	E_a^*, kJ·mol⁻¹
Blank		45	323			
	0	55	340	1.8	−191.76	4.40
		25	24			
Itraconazole		45	161			
	15	55	213	59.1	−19.33	61.75
		25	44			
Fluconazole	15	45	216	51.2	−41.07	53.84
		55	301			

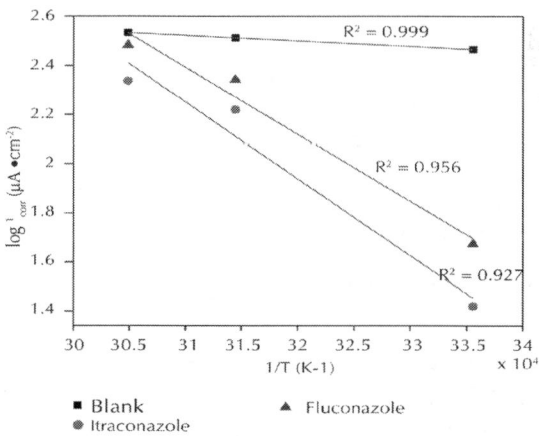

Figure 9: Arrhenius plots in the absence and presence of inhibitors.

From the results, the activation energy value in the presence of Itraconazole is higher than that of Fluconazole, which support the previous results that Itraconazole has higher efficiency than Fluconazole.

The enthalpy (ΔH^*) and entropy (ΔS^*) of activation was calculated from the following equation:

$$K = \frac{RT}{Nh\exp\left(\Delta S^* / R\right)\exp\left(-\Delta H^* / RT\right)}$$

(7)

where h is Planck's constant = 6.626×10^{-34} Js and is Avogadro's number = 6.022×10^{23} mol^{-1}. Figure 10shows the plots utilized to get the above parameters which are listed in Table 7. The increase of ΔH^* values in the presence of the two inhibitors may be attributed to the increased heights of the energy barriers of the corrosion process. As ΔH^* values are lower than E_a^*, the corrosion process must involve a gaseous reaction, simply the hydrogen evolution reaction, associated with a decrease in the total reaction volume. This result verified the known thermodynamic relation between E_a^* and ΔH^* in the following equation [32, 36]:

$$\Delta H^* = E_a^* - RT$$

(8)

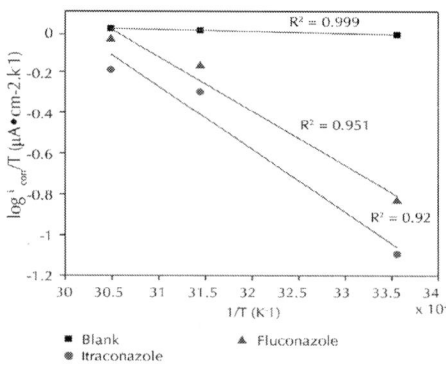

Figure 10: Variation of log i_{corr} / T versus 1/ T in the absence and presence of inhibitors.

As shown also from Table 7, ΔS^* values are large and negative both in the absence and presence of the two inhibitors, implying that the activated complex represented the rate determining step with respect to the association rather than dissociation. It means that a decrease in disorder occurred when proceeding from reactants to the activated complex [18, 37]. In addition, the less negative values of ΔS^* in the presence of inhibitors create a near-equilibrium corrosion system state [36].

As observed from the above kinetic activation parameters, Itraconazole compound exhibit higher inhibition effect than Fluconazole compound on the API 5L-B in the 3.5% NaCl solution saturated with CO_2.

SEM, EDX, and XRD Characterization

SEM images were taken and EDX/XRD analysis was performed in the absence and presence of 15 ppm Itraconazole and Fluconazole for three days. As shown in SEM images (Figure 11), corrosion signs appeared on the metal surface in the uninhibited solution, Figure 11(a), and nearly disappeared in the metal surface of the inhibited solutions, Figures 11(b) and 11(c).

2 mm ⟶ x35

(a)

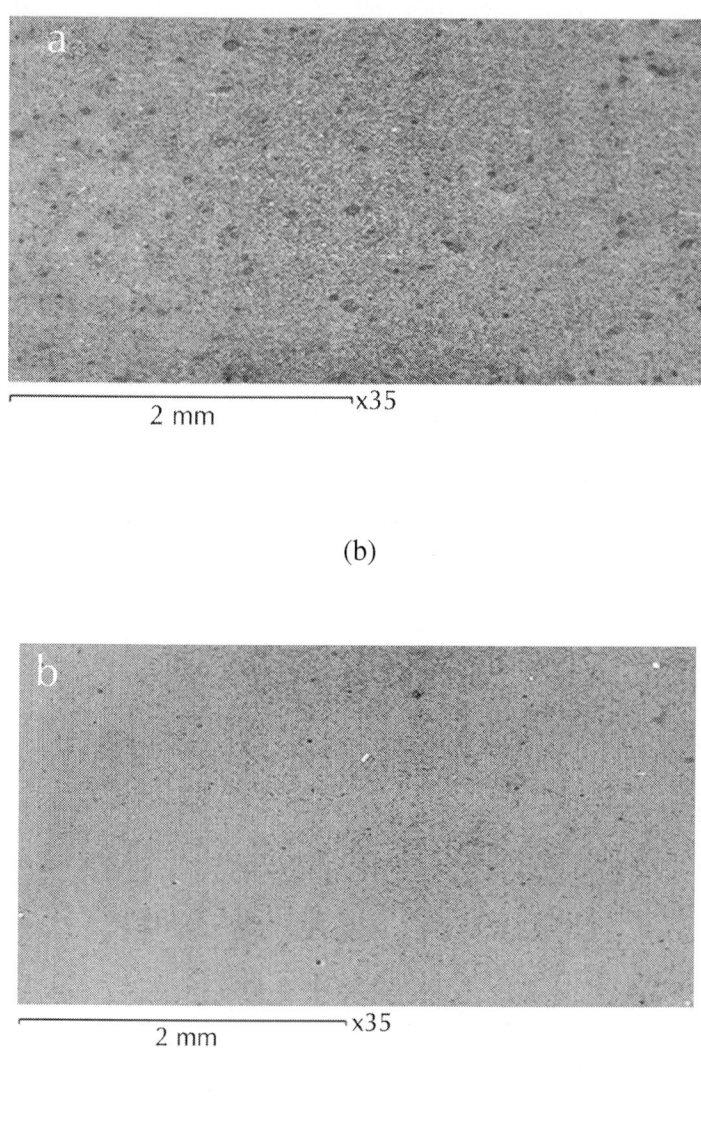

(b)

(c)

Figure 11: SEM images in the absence (a) in the presence of Itraconazole (b) and Fluconazole (c).

Figures 12(a)–12(c) show the EDX spectrum in the absence and presence of Itraconazole and Fluconazole. In the presence of inhibitors, Figures 12(b) and 12(c), EDX spectra show an additional line characteristic for the existence of N. In addition, the intensities of C, and O signals are enhanced. The appearance of the N signal and this enhancement in the C and O signals is due to the N, C and O atoms constituting the inhibitors compounds which indicate that the inhibitor molecules have adsorbed on the metal surface. Data obtained from the spectra are presented in Tables 8(a)–8(c). The spectra show also that Fe peaks are considerably suppressed in the presence of inhibitor which is due to the overlying inhibitor film. These results confirm those from electrochemical measurements which suggest that a surface film inhibits the metal dissolution, and hence retarded the hydrogen evolution reaction [38, 39].

Table 8: (a) Data obtained from EDX in the absence of inhibitors. (b) Data obtained from EDX in the presence of Itraconazole. (c) Data obtained from EDX in the presence of Fluconazole

(a)

Element	App Conc.	Intensity Corrn.	Weight%	Weight% Sigma	Atomic%
C K	23.46	0.3861	7.41	0.66	25.34
O K	26.22	0.9789	3.27	0.31	18.39
Na K	0.94	0.2435	0.47	0.21	0.84
Cl K	0.96	0.6974	0.17	0.06	0.19
Mn K	4.03	0.9508	0.52	0.09	0.39
Fe K	705.48	0.9756	88.17	0.72	64.85
Totals			100.00		

(b)

Element	App Conc.	Intensity Corrn.	Weight%	Weight% Sigma	Atomic%
C K	59.48	0.4193	13.66	0.65	31.68
N K	1.09	0.1868	0.56	0.63	1.12
O K	174.71	0.8666	19.41	0.42	33.80

Cl K	7.93	0.7235	1.06	0.06	0.83
Mn K	2.50	0.9092	0.26	0.07	0.13
Fe K	628.41	0.9303	65.04	0.71	32.44
Totals			100.00		

(c)

Element	App Conc.	Intensity Corrn.	Weight%	Weight% Sigma	Atomic%
C K	49.25	0.3988	12.85	0.68	34.16
N K	0.80	0.1860	0.45	0.61	1.02
O K	69.13	0.8586	8.38	0.35	16.72
Na K	10.04	0.2682	3.89	0.23	5.41
Cl K	3.09	0.7080	0.45	0.06	0.41
Mn K	2.92	0.9260	0.33	0.07	0.19
Fe K	671.46	0.9486	73.65	0.80	42.10
Totals			100.00		

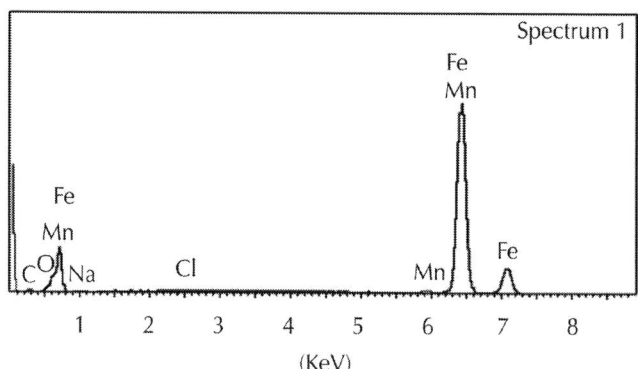

Full scale 16313 cts cursor: 8.885 (45cts)

(a)

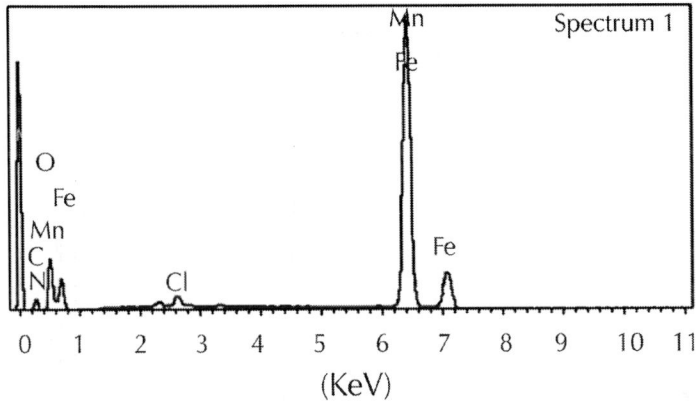

Full scale 10538 cts cursor: 11.082 (40cts)

(b)

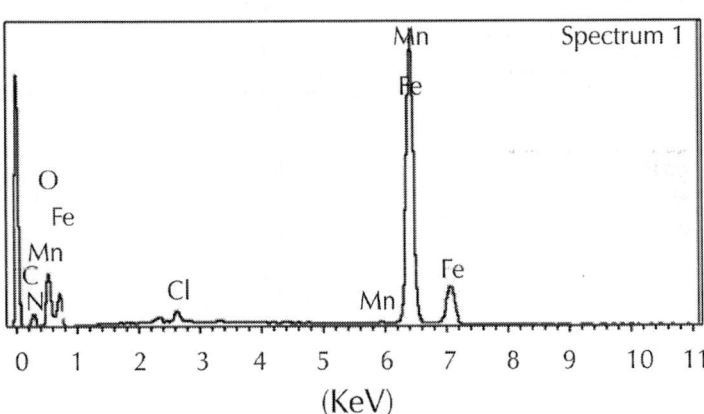

Full scale 10538 cts cursor: 11.082 (40cts)

(c)

Figure 12: EDX spectra in the absence (a) and in the presence of Itraconazole (b) and Fluconazole (c).

Figures 13(a)–13(c) shows the XRD pattern in the absence and presence of Itraconazole and Fluconazole. In the absence of inhibitors, Figure 13(a), corrosion products as $(H_3O)_2FeCl_5(H_2O)$ and $Fe_2O_3 \cdot H_2O$ have been observed on the metal surface. Intensity of these compounds is much higher on the metal surface as indicated by the related peaks. In the presence of inhibitors, Figures 13(b) and 13(c), iron nitrides as Fe_2N and Fe_8N have been observed on the metal surfaces which represent the protective films formed due to interaction of iron with the inhibitor molecules. Intensity of these compounds is much higher on the metal surface as indicated by the related peaks which indicate that these compounds have covered the metal surface and protected iron from the corrosive media. Also, narrow peaks of these compounds indicate that the formed films have crystallized characteristics; this led to increase in the level of protection which is confirmed by the absence of corrosion products in the presence of iron nitrides [40]. Nitride coatings have been used in numerous applications to increase the hardness and improve the wear and corrosion resistance of structural materials, as well as in various high-tech areas, where their functional rather than mechanical properties are of prime importance [41, 42].

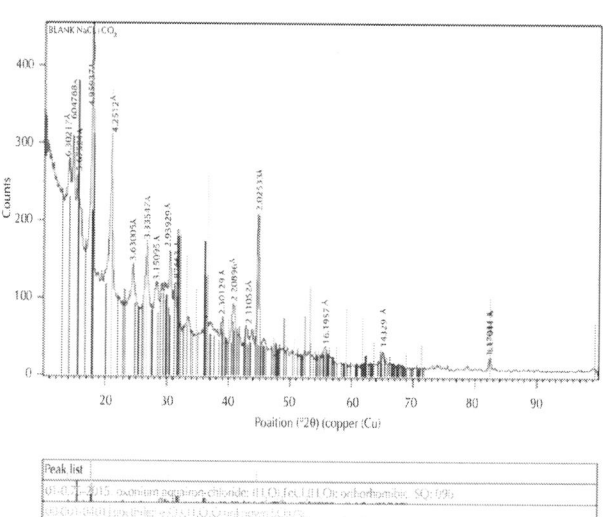

(a)

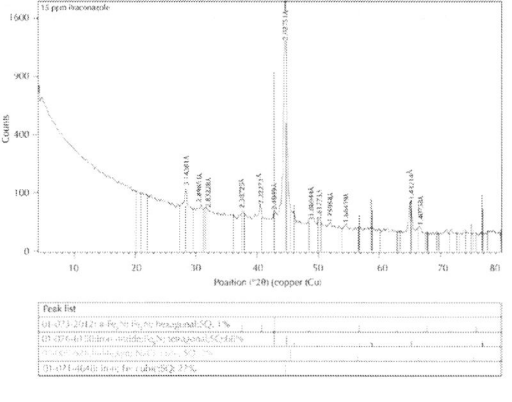

(b)

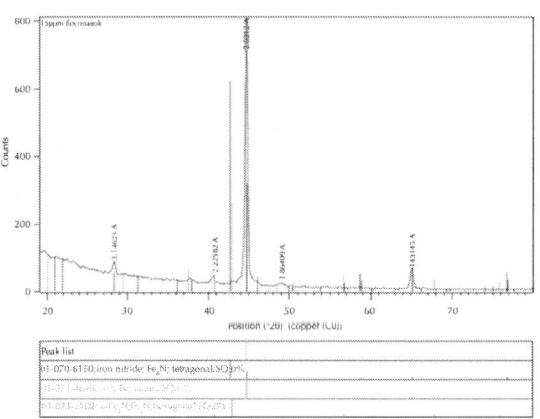

(c)

Figure 13: (a) XRD pattern for the API 5L-B specimen after immersion in 3.5% NaCl solution saturated with CO_2 for 3 days. (b) XRD pattern for the API 5L-B specimen after immersion in 3.5% NaCl solution saturated with CO_2 with 15 ppm Itraconazole for 3 days. (c) XRD pattern for the API 5L-B specimen after immersion in 3.5% NaCl solution saturated with CO_2 with 15 ppm Fluconazole for 3 days.

MECHANISM OF INHIBITION

Two modes of adsorption are considered on the metal surface in acid media. In the first mode, the neutral molecules may be adsorbed on the surface of carbon steel through the chemisorption mechanism, involving the displacement of water molecules from the carbon steel surface and sharing the electrons between the hetero-atoms and iron. The inhibitor molecules can also adsorb on the steel surface on the basis of donor-acceptor interactions between ϖ-electrons of aromatic ring and vacant d-orbitals of surface iron atoms. In the second mode, since it is well known that the steel surface bears positive charge in acid solution [43, 44], so it is difficult for the protonated molecules to approach the positively charged steel surface due to the electrostatic repulsion. Since chloride ions have a smaller degree of hydration, thus and they could bring excess negative charges in the vicinity of the interface and favor more adsorption of the positively charged inhibitor molecules, the protonated investigated compounds adsorb through electrostatic interactions between the positively charged molecules and negatively charged metal surface. Thus, there is a synergism between adsorbed Cl^- ions and protonated investigated compounds. Thus we can conclude that inhibition of API 5L-B corrosion in CO_2-saturated 3.5% NaCl solutions is mainly due to electrostatic interaction. The decrease in inhibition efficiency with rise in temperature supports electrostatic interaction.

The higher inhibition of Itraconazole in comparison with Fluconazole may be attributed to the larger number of nitrogen atoms in the molecular structure of the former. In addition, acid anions such as Cl^- and F^- may be specifically adsorbed on the metal surface, donating an excess negative charge to the metal surface. In this way, potential of zero charge (PZC) becomes less negative which promotes the adsorption of inhibitors in cationic form. These observations were addressed by others [28, 32, 45].

CONCLUSIONS

Itraconazole and Fluconazole compounds are good inhibitors for API 5L-B in 3.5% NaCl solution saturated with CO_2. The two compounds classified as mixed type as their molecules adsorbed on the metal surface, forming protective film, and blocking the available reaction sites exposed to the corrosive medium. At 25°C, corrosion rates decrease by increasing inhibitor concentrations to reach the lowest values at 15 and 20 ppm for Itraconazole and Fluconazole, respectively. Inhibition efficiency reached to 92% and 90% for the two inhibitors at these concentrations. At higher temperatures, corrosion rates increased in the absence and presence of the two inhibitors.

The two inhibitors adsorbed on the steel surface according to Langmuir isotherm. ΔG_{ads}° values are large and negative which means that the process is spontaneous. ΔG_{ads}° values are higher than 40 kJ·mol^{-1} which means that the adsorption is chemisorption.

E_a^* values increased by the addition of the two compounds, so that the activation energy of the metal dissolution increased due to the adsorption of the organic molecules on the metal surface, forming a double layer which is considered a barrier for corrosion. ΔH^* values are lower than indicating that the corrosion process must involve a gaseous reaction, simply the hydrogen evolution reaction, associated with a decrease in the total reaction volume. ΔS^* values are large and negative in the absence and presence of the two inhibitors, implying that the activated complex represented the rate determining step with respect to the association rather than dissociation. It means that a decrease in disorder occurred when proceeding from reactants to the activated complex.

SEM images show corrosion as scattered pits on the metal surface immersed in the test solution without inhibitors, while those of the specimens after adding 15 ppm of inhibitors show clean surfaces and disappear of most corrosion signs. EDX spectra for specimens inserted in the inhibited solutions showed an additional line characteristic for the existence of N. Also, intensities of C and O signals are enhanced. These results confirm that a carbonaceous material containing these atoms has covered the metal surface. This layer is undoubtedly due to the inhibitor. Also, Fe peaks in case of inhibited solutions are suppressed due to the overlying inhibitor film. These results support that a surface film has formed and reduced the metal dissolution. XRD patterns for specimen surface in case of inhabited solutions show the formation of iron nitrides (Fe_2N and Fe_8N) on the metal surface which represent the protective film formed due to interaction of iron with the inhibitor molecules.

All above results show that Itraconazole and Fluconazole compounds are good inhibitors for API 5L-B in 3.5% NaCl solution saturated with CO_2. Also, all results show that Itraconazole compound has higher inhibition efficiency than Fluconazole despite its lower cost.

REFERENCES

1.   P. C. Okafor, X. Liu, and Y. G. Zheng, "Corrosion inhibition of mild steel by ethylamino imidazoline derivative in CO_2-saturated solution," Corrosion Science, vol. 51, no. 4, pp. 761–768, 2009.

2.   M. Heydari and M. Javidi, "Corrosion inhibition and adsorption behaviour of an amido-imidazoline derivative on API 5L X52 steel in CO_2-saturated solution and synergistic effect of iodide ions,"Corrosion Science, vol. 61, pp. 148–155, 2012.

3.   X. Liu, P. C. Okafor, and Y. G. Zheng, "The inhibition of CO_2 corrosion of N80 mild steel in single liquid phase and liquid/ particle two-phase flow by aminoethyl imidazoline derivatives," Corrosion Science, vol. 51, no. 4, pp. 711 751, 2009.

4.   B. D. Mert, M. Erman Mert, G. Karda , and B. Yazıc, "Experimental and theoretical investigation of 3-amino-1, 2, 4-triazole-5-thiol as a corrosion inhibitor for carbon steel in HCl medium," Corrosion Science, vol. 53, no. 12, pp. 4265–4272, 2011.

5.   B. S. Sanatkumar, J. Nayak, and A. N. Shetty, "Influence of 2-(4-chlorophenyl)-2-oxoethyl benzoate on the hydrogen evolution and corrosion inhibition of 18 Ni 250 grade weld aged maraging steel in 1.0 M sulfuric acid medium," International Journal of Hydrogen Energy, vol. 37, pp. 9431–9442, 2012.

6.   W. H. Li, Q. He, S. T. Zhang, C. L. Pei, and B. R. Hou, "Some new triazole derivatives as inhibitors for mild steel corrosion in acidic medium," Journal of Applied Electrochemistry, vol. 38, no. 3, pp. 289–295, 2008.

7.   S. John, K. M. Ali, and A. Joseph, "Electrochemical, surface analytical and quantum chemical studies on Schiff bases of 4-amino-4H-1,2,4-triazole-3,5-dimethanol (ATD) in corrosion

protection of aluminium in 1N HNO$_3$," Journal of Materials Science, vol. 34, no. 6, pp. 1245–1256, 2011.

8. S. John, J. Joy, M. Prajila, and A. Joseph, "Electrochemical, quantum chemical, and molecular dynamics studies on the interaction of 4-amino-4H,3,5-di(methoxy)-1,2,4-triazole (ATD), BATD, and DBATD on copper metal in 1N H$_2$SO$_4$," Materials and Corrosion, vol. 62, no. 11, 2011.

9. F. A. Ansari and M. A. Quraishi, "Inhibitive effect of some gemini surfactants as corrosion inhibitors for mild steel in acetic acid media," Arabian Journal for Science and Engineering, vol. 36, no. 1, pp. 11–20, 2011.

10. F. Bentiss, M. Traisnel, and M. Lagrenee, "The substituted 1,3,4-oxadiazoles: a new class of corrosion inhibitors of mild steel in acidic media," Corrosion Science, vol. 42, no. 1, pp. 127–146, 2000.

11. S. Muralidharan, K. L. N. Phani, S. Pitchumani, S. Ravichandran, and S. V. K. Iyer, "Polyamino-benzoquinone polymers: a new class of corrosion inhibitors for mild steel," Journal of the Electrochemical Society, vol. 142, no. 5, pp. 1478–1483, 1995.

12. M. G. Hosseini, M. Ehteshamzadeh, and T. Shahrabi, "Protection of mild steel corrosion with Schiff bases in 0.5 M H$_2$SO$_4$ solution," Electrochimica Acta, vol. 52, no. 11, pp. 3680–3685, 2007.

13. H. Wang, X. Wang, H. Wang, L. Wang, and A. Liu, "DFT study of new bipyrazole derivatives and their potential activity as corrosion inhibitors," Journal of Molecular Modeling, vol. 13, no. 1, pp. 147–153, 2007.

14. Gamry Echem Analyst Manual, 2003.

15. R. W. Bosch, "Electrochemical frequency modulation: a new electrochemical technique for online corrosion monitoring," Corrosion, vol. 57, no. 1, pp. 60–70, 2001.

16. Y. M. Tang, Y. Chen, W. Z. Yang, Y. Liu, X. S. Yin, and J. T. Wang, "Electrochemical and theoretical studies of thienyl-substituted amino triazoles on corrosion inhibition of copper in 0.5 M H$_2$SO$_4$,"Journal of Applied Electrochemistry, vol. 38, no. 11, pp. 1553–1559, 2008.

17. M. A. Migahed, "Electrochemical investigation of the corrosion behaviour of mild steel in 2 M HCl solution in presence of

1-dodecyl-4-methoxy pyridinium bromide," Materials Chemistry and Physics, vol. 93, no. 1, pp. 48–53, 2005.

18. A. S. Fouda, G. Y. Elewady, and M. N. El-Haddad, "Corrosion inhibition of carbon steel in acidic solution using some azodyes," Canadian Journal on Scientific and Industrial Research, vol. 2, no. 1, 2011.

19. H. Amar, A. Tounsi, A. Makayssi, A. Derja, J. Benzakour, and A. Outzourhit, "Corrosion inhibition of Armco iron by 2-mercaptobenzimidazole in sodium chloride 3% media," Corrosion Science, vol. 49, no. 7, pp. 2936–2945, 2007.

20. G. Avci, "Corrosion inhibition of indole-3-acetic acid on mild steel in 0.5 M HCl," Colloids and Surfaces A, vol. 317, no. 1-3, pp. 730–736, 2008.

21. M. Abdallah, "Rhodanine azosulpha drugs as corrosion inhibitors for corrosion of 304 stainless steel in hydrochloric acid solution," Corrosion Science, vol. 44, no. 4, pp. 717–728, 2002.

22. M. Lebrini, M. Lagrenée, M. Traisnel, L. Gengembre, H. Vezin, and F. Bentiss, "Enhanced corrosion resistance of mild steel in normal sulfuric acid medium by 2,5-bis(n-thienyl)-1,3,4-thiadiazoles: electrochemical, X-ray photoelectron spectroscopy and theoretical studies," Applied Surface Science, vol. 253, no. 23, pp. 9267–9276, 2007.

23. G. Moretti, F. Guidi, and G. Grion, "Tryptamine as a green iron corrosion inhibitor in 0.5 M deaerated sulphuric acid," Corrosion Science, vol. 46, no. 2, pp. 387–403, 2004.

24. F. M. Donahue and K. Nobe, "Theory of organic corrosion inhibitors adsorption and linear free energy relationships," Journal of The Electrochemical Society, vol. 112, no. 9, pp. 886–891, 1965.

25. E. Kamis, F. Belluci, R. M. Latanision, and E. S. H. El-Ashry, "Acid corrosion inhibition of nickel by 2-(triphenosphoranylidene) succinic anhydride," Corrosion, vol. 47, no. 9, pp. 677–686, 1991.

26. O. Benali, L. Larabi, M. Traisnel, L. Gengembre, and Y. Harek, "Electrochemical, theoretical and XPS studies of 2-mercapto-1-methylimidazole adsorption on carbon steel in 1 M HClO$_4$," Applied Surface Science, vol. 253, no. 14, pp. 6130–6139, 2007.

27. R. Solmaz, G. Karda, M. Çulha, B. Yazici, and M. Erbil, "Investigation of adsorption and inhibitive effect of 2-mercaptothiazoline on corrosion of mild steel in hydrochloric acid media," Electrochimica Acta, vol. 53, no. 20, pp. 5941–5952, 2008.

28. J. Aljourani, K. Raeissi, and M. A. Golozar, "Benzimidazole and its derivatives as corrosion inhibitors for mild steel in 1 M HCl solution," Corrosion Science, vol. 51, no. 8, pp. 1836–1843, 2009.

29. F. Bentiss, M. Lebrini, and M. Lagrenée, "Thermodynamic characterization of metal dissolution and inhibitor adsorption processes in mild steel/2,5-bis(n-thienyl)-1,3,4-thiadiazoles/ hydrochloric acid system," Corrosion Science, vol. 47, no. 12, pp. 2915–2931, 2005.

30. A. Y. Musa, A. B. Mohamad, A. A. H. Kadhum, M. S. Takriff, and L. T. Tien, "Synergistic effect of potassium iodide with phthalazone on the corrosion inhibition of mild steel in 1.0 M HCl," Corrosion Science, vol. 53, no. 11, pp. 3672–3677, 2011.

31. S. Ghareba and S. Omanovic, "The effect of electrolyte flow on the performance of 12-aminododecanoic acid as a carbon steel corrosion inhibitor in CO_2-saturated hydrochloric acid," Corrosion Science, vol. 53, no. 11, pp. 3805–3812, 2011.

32. E. A. Noor and A. H. Al-Moubaraki, "Thermodynamic study of metal corrosion and inhibitor adsorption processes in mild steel/1-methyl-4[4'(-X)-styryl pyridinium iodides/hydrochloric acid systems," Materials Chemistry and Physics, vol. 110, no. 1, pp. 145–154, 2008.

33. L. Larabi, Y. Harek, O. Benali, and S. Ghalem, "Hydrazide derivatives as corrosion inhibitors for mild steel in 1 M HCl," Progress in Organic Coatings, vol. 54, no. 3, pp. 256–262, 2005.

34. Y. P. Khodyrev, E. S. Batyeva, E. K. Badeeva, E. V. Platova, L. Tiwari, and O. G. Sinyashin, "The inhibition action of ammonium salts of O,O-dialkyldithiophosphoric acid on carbon dioxide corrosion of mild steel," Corrosion Science, vol. 53, no. 3, pp. 976–983, 2011.

35. K. K. Al-Neami, A. K. Mohamed, I. M. Kenawy, and A. S. Fouda, "Inhibition of the corrosion of iron by oxygen and nitrogen

containing compounds," Monatshefte für Chemie Chemical Monthly, vol. 126, no. 4, pp. 369–376, 1995.

36. Y. A. Elewady, A. S. Fouda, and H. K. Zeid, "Corrosion behavior of some petroleum equipment in acidic media," Physical Chemistry, Mansoura University. In press.

37. G. E. Badr, "The role of some thiosemicarbazide derivatives as corrosion inhibitors for C-steel in acidic media," Corrosion Science, vol. 51, no. 11, pp. 2529–2536, 2009.

38. S. S. A. Rehim, O. A. Hazzazi, M. A. Amin, and K. F. Khaled, "On the corrosion inhibition of low carbon steel in concentrated sulphuric acid solutions—part I: chemical and electrochemical (AC and DC) studies," Corrosion Science, vol. 50, no. 8, pp. 2258–2271, 2008.

39. L. Fragoza-Mar, O. Olivares-Xometl, M. A. Domnguez-Aguilar, E. A. Flores, P. Arellanes-Lozada, and F. Jiménez-Cruz, "Corrosion inhibitor activity of 1, 3-diketone malonates for mild steel in aqueous hydrochloric acid solution," Corrosion Science, vol. 61, pp. 171–184, 2012.

40. P. S. Prevéy, "X-ray diffraction characterization of crystallinity and phase composition in plasma-sprayed hydroxyapatite coatings," Journal of Thermal Spray Technology, vol. 9, no. 3, pp. 369–376, 2000.

41. I. Bertóti, "Characterization of nitride coatings by XPS," Surface and Coatings Technology, vol. 151-152, pp. 194–203, 2002.

42. J. Baranowska and S. E. Franklin, "Characterization of gas-nitrided austenitic steel with an amorphous/nanocrystalline top layer," Wear, vol. 264, no. 9-10, pp. 899–903, 2008.

43. G. N. Mu, T. P. Zhao, M. Liu, and T. Gu, "Effect of metallic cations on corrosion inhibition of an anionic surfactant for mild steel," Corrosion, vol. 52, no. 11, pp. 853–856, 1996.

44. A. K. Singh and M. A. Quraishi, "Inhibitive effect of diethylcarbamazine on the corrosion of mild steel in hydrochloric acid," Corrosion Science, vol. 52, no. 4, pp. 1529–1535, 2010.

45. A. Popova, M. Christov, S. Raicheva, and E. Sokolova, "Adsorption and inhibitive properties of benzimidazole derivatives in acid mild steel corrosion," Corrosion Science, vol. 46, no. 6, pp. 1333–1350, 2004.

Corrosion Control in Industry

B. Valdez[1], M. Schorr[1], R. Zlatev[1], M. Carrillo[1], M. Stoytcheva[1], L. Alvarez[1], A. Eliezer[2], nd N. Rosas[3]

[1]Instituto de Ingeniería, Departamento de Materiales, Minerales y Corrosión, Universidad Autónoma de Baja California, Mexicali, Baja California, México

[2]Sami Shamoon College of Engineering Corrosion Research Center, Ber Sheva, Israel

[3]Unversidad Politécnica de Baja California, Mexicali, Baja California, México

INTRODUCTION

The economic development of any region, state or country, depends not only on its natural resources and productive activities, but also

on the infrastructure that account for the exploitation, processing and marketing of goods. Irrigation systems, roads, bridges, airports, maritime, land and air transport, school buildings, offices and housing, industrial installations are affected by corrosion and therefore susceptible to deterioration and degradation processes.

Corrosion is a worldwide crucial problem that strongly affects natural and industrial environments. Today, it is generally accepted that corrosion and pollution are interrelated harmful processes since many pollutants accelerate corrosion and corrosion products such as rust, also pollute water bodies. Both are pernicious processes that impair the quality of the environment, the efficiency of the industry and the durability of the infrastructure assets. Therefore, it is essential to develop and apply corrosion engineering control methods and techniques.

Other critical problems, that impact on infrastructure and industry are climate change, global warming and greenhouse emissions, all interrelated phenomena.

This chapter presents important aspects of corrosion in industrial infrastructure, its causes, impacts, control, protection and prevention methods.

MATERIALS IN INDUSTRY

Metallic materials play a key role in the development of a country and its sustained growth in the context of the global economy. Table 1 shows a classification and the properties of different types of materials used in the industry. During the course of the metal production it undergoes various types of processes: mining of minerals, manufacturing and application and generation of gases, liquids or solids that are released into the environment. In the industrial development, production and use of materials in general, economic cycles are due to take effect that influence the environment (Raichev et al., 2010). The selection of a predominant group of materials depends on the particular industries; they determine to a greater or lesser extent the pattern of consumption of a given product, inducing the market to adapt itself to this new reality. The materials industry follows two general strategies: research the materials and the available technology recommended for their. Recycled materials typically require less capital and energy consumption, but need more manpower, for primary processing.

Also, the costs of pollution control are lower than those required for primary processing of minerals. Recycling becomes more intense, as economies tend to be more sophisticated, since viable quantities of recycled material must be available for reuse (Garcia, R., et al, 2012, Lopez, G. 2011, Schorr, M., 2010).

Table 1: Materials in industry: Types, main properties and uses

Material	Main properties	Uses
Metals and alloys (carbon and stainless steels, non ferrous alloys)	Mechanical resistance hardness	Cars, aircraft, tanks, infrastructure reinforcement.
Plastics (Synthetic polymers, rubbers)	Low density and corrosion resistance	Process components, tubes, vessels, coatings, paints.
Ceramics (Metallic carbides, silica, glass, alumina)	High hardness, high temperature and corrosion resistance	Cutting tools, motor components, refractory bricks, ovens, etc.
Composites (glass and carbon fibers reinforced plastics, plastic matrixes reinforced with metallic particles)	Light weight, high strength and hardness.	Car bodies, aircraft components, vessels, construction.

In the production of a material waste is generated: for example, parts of material that was left aside, through the production steps. There are called effluent, which consist of waste that comes from the processes linked to the technology involved in each step of production, although not necessarily with the main material. Industrial processes for the recovery of ore from the mine to produce a metal, are related to technological development and therefore varies from one country to another, including regulatory laws, financial aspects etc.. Therefore, the environmental impacts vary widely. A low grade or poor quality of the ore, with low metal content, increase the cost of recovery, requiring large amounts of mineral raw material and energy invested for the recovery of small amounts of metal. Also important is the feasibility of the mineral that can be worked out e.g., the cost of physical removal of rock, accessibility to the mines, thickness and regularity of the ore zone, and its hardness. Figure 1, shows the material cycle, which involves processes from raw material, extraction from natural sources, processing and conversion into industrial materials, their processing and application, the deterioration rate effects, its mechanical properties, environmental behavior, corrosion, disposal and possible recovery of some of these through the use of recycling methods.

There are many examples of recovery of metals, which could help to describe step by step the various interactions with the environment itself. A mineral submitted to a production process will impact the environment, during four steps: extraction, processing, fabrication and manufacturing, of goods as seen in the cycle of materials. (Figure 1).

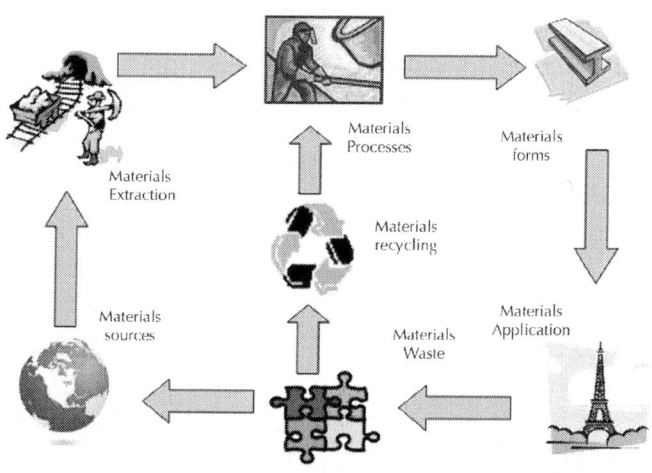

Figure 1: Materials production and use cycle.

In the mineral extraction step, the effluents of N, C, S, NO_x, SO_x and CO_x, from machinery and equipment, operation process water, particulate matter and ground movement in landfills.

The processing stage, chemical operations or extractive metallurgy for converting the concentrate into metal apply selected technologies. The effluents are gases such as SO_2, NO_2 and CO_2, water contaminated with heavy metals, and hazard sediments.

In the manufacturing step the material undergoes operations that transform it into rods, bars, sheets; losses are scrap metal, such as cuts, burrs, mill scale, which recycled with no net loss of metal. In the manufacturing stage the metal is formed by stamping, machining and forging.

Focus on good operations management involves control of air emissions, water management and treatment, solid waste disposal and good land use, will greatly help to maintain a good balance with the environment. It is also necessary to analyze the production area to identify what improvements or measures should be implemented. The role of hydrometalurgist is particularly important and so he is responsible for the design of environmentally friendly processes in each of his steps, to promote sustainable production.

Processes of Materials Biodeterioration in Industrial Systems

In addition to the common processes of deterioration of materials by chemical reactions and mechanical fracture, there are others who are concerned with the participation of various types of microorganisms that adhere in colonies or develop on their surfaces.

Biocorrosion and biodeterioration of metallic materials and nonmetallic materials are two important processes that cause serious problems to the infrastructure of various industrial systems. Generally, microorganisms do not deteriorate or corrode metals directly, but modify the conditions of interface material / environment and surroundings, favoring the degradation of these materials in such a way that induce or influence the development process.

Biofouling is a common term that indicates the presence of microbiological growth on the surfaces of structures built of different materials favoring the formation of biofilms with the colonies of various types of microorganisms.

In the case of metal, biocorrosion occurs due to corrosion electrochemical processes and biological agents due to the action of microorganisms and / or bacteria present in the system. The knowledge of these biological processes and their effects is necessary in order to establish preventive measures and control measures in industrial systems.

An industrial plant containing several biocorrosion environments is a potential risk:

In a heat exchangers system, usually dust accumulates biological waste; biocorrosion could occur, leading to corrosion film formation on walls surface. Therefore, it will be energy loss by increasing the resistance to fluid flow and heat transfer. Loss by evaporation of water favors the increase of the concentration of nutrients, the residence time, the water temperature and the surface / volume ratio, which leads to higher rate of microbial growth (Stoytcheva et al., 2010, Carrillo M. et al., 2010).

Until the early 80's of the twentieth century, we used mixtures of anodic and cathodic inhibitors, such as chromium, zinc and phosphates, to lessen the effects of corrosion in water systems. In some cases we added a polymer, as is still done to date, to avoid or eliminate the problems of fouling on the metal walls. On the other hand, to prevent microbiological growth, we added biocides such as chlorine and quaternary ammonium compounds under acidic conditions.

In the early 90's, the strategies for industrial water treatment changed because of pressure from laws inforcing for the preservation of the environment. Chromates and acid pH values are replaced by the use of organic phosphonates as corrosion inhibitors, while for the control of fouling polycarboxylate type polymers are used. However, this change brought about an increase in the amount of suspended solids, a greater number and variety of microorganisms and therefore a greater amount of inorganic deposits on the heat exchangers walls.

Biodeterioration of Metallic and Nonmetallic Materials

The metal nature has an effect on the distribution and development of microbial films on its surface. These films influence on the wear and corrosion of the metal substrate. The lack of homogeneity in the biofilm is a precursor of differential aeration processes with formation of differential cell concentration, for example, stainless steels (SS) and nickel-copper (Ni-Cu) alloys in seawater. The oxides passive films or hydrated hydroxides (corrosion products) are a good place for the establishment and growth of bacteria, especially when these products are at a physiological pH values (pH $\approx$ 7.4)

Carbon Steel (CS)

CS are very active metals in aggressive media, such as seawater. In this case, the action of microorganisms involves the dissolution of films of corrosion products, by processes of oxidation and reduction. This creates new metal active areas, exposed to the aggressive medium and suffers corrosion processes. In the case of sulfate-reducing bacteria (SRB), the species generated by their metabolism (sulfides) are corrosive to the metal. Figure 2 shows the final state pitting outside a CS pipe, which was affected by microbial growth inside, prompting a process of microbial corrosion with not uniform localized attack.

Stainless Steel

The presence of chromium and molybdenum as alloying elements, enable passive behavior of stainless steels in different environments. However, the passive surface of these SS provides an ideal location for microbial adhesion and therefore are susceptible to corrosion pitting, crevice corrosion under stress or in solutions containing chlorides, as sea water.

Figure 2: External pitting caused by biocorrosion on the internal surface of a carbon steel pipe in a fire extinguisher system.

In marine environments, the generation of peroxides during bacterial metabolism causes an ennoblement of the pitting potential of SS, thus promoting corrosion. Obviously, not all SS have the same behavior, but in general they tend to deteriorated in the presence of colonies of microorganisms.

Copper and Nickel Alloys

Alloys of Cu with Zn, Sn and Al, brasses, bronzes, aluminum bronzes; also the nickel alloys: Monel, Hastelloy, nickel superalloys: Ni-Mo, Ni-Cr-Mo, Ni-Cr-Fe- Mo; the traditional nickel alloys: Ni-Cr-Fe, Ni-Fe-Cr, Fe-Ni-Cr-Mo), and the Cu-ni alloys CuNi\70/30, CuNi\90/10, have shown great corrosion resistance in different environments, so they have found a wide use in different industries and environments. However, despite these skills, there are reports that these alloys are colonized by bacteria after several months of exposure in seawater (Acuña, N. et al., 2004).

Aluminum and its Alloys

Al is an active metal which is passivated rapidly in some neutral and acid media, thus offering a good resistance to corrosion. Al alloys with copper, magnesium and zinc, are widely used in the aviation industry. However, there have been cases of biocorrosion on fuel tanks of jet aircraft made of Al alloys by microbial contaminants in turbo combustibles. The presence of water (moisture), even in minimal amounts, allows growth of microorganisms (typically fungi), when these are able to utilize hydrocarbons as a carbon source.

Titanium

Ti is considered as the most resistant metal to biocorrosion, according to the results of tests carried in different conditions, due to its passive behavior that is reinforced in the presence of oxidizing agents. This is the reason why Ti is the material of choice, for example, for the manufacture of tubes in cooling systems that use seawater.

Nonmetallic Materials

Non-metallic materials such as fiberglass reinforced polyester (FGRP), concrete and wood, are also affected by biodeterioration processes in the presence of microorganisms

In the case of FGRP, bacteria and algae are able to use the polyester matrix as a carbon source, consuming and considerably reducing the mechanical strength of composite material, ultimately causing its failure. This is easily observable in screens of this material in cooling towers or tanks containing fresh water or salt water. Wood suffers biodeterioration by the presence of fungi in moist environments that promote the delignification of this material (Valdez B., et al., 1996, 1999, 2008).

Facing the Problems of Biodeterioration

The inevitable presence of microorganisms in the feed water causes a sequence of biofouling, biocorrosion and biodeterioration of the materials component of the structures. This sequence depends on the degree of microbial contamination and the system operating characteristics.

The most common methods of controlling these problems involve the application of continuous or metered biocides such as chlorine. Currently, we use substances more compatible with the environment, since the use of chlorine is limited to certain concentrations. Such is the case of ozone, which is also ascribed with passivating effects on certain metals and alloys commonly applied in industry, and also in antifouling action.

In order to tackle a biodeterioration problem it is required a prior analysis of the problem, to know when conditions are suitable for the development of this process. In industrial systems we need to know some parameters: temperature, pH, nutrients; carbon, phosphorus, nitrogen, sulfate ion levels and flow rates. The places where we find biodeterioration are: biofouling deposits, under any deposit, zones of localized metal corrosion. To check their presence it is necessary to utilize sampling techniques, isolation and identification of microorganisms. It is interesting to note that there are commercial devices for in situ measurements that are practical and useful for the plant engineer.

CORROSION IN THE ELECTRONICS INDUSTRY

Corrosion of device components, manufactured by the electronics industry, is a problem that has occurred during a long time. Often, especially corrosion of one or more of the metallic elements of an electronic component is the primary cause of failure in various electronic equipments. The high density of components required to reduce the size of electronic equipment, also for a better signal processing, leads to the generation of enclosed corrosion between thin metal sections. Furthermore, when electronic devices are in more severe environments such as tropical, subtropical, contaminated deserts, etc., they have high failure rates. Problems, due to the aggressiveness of the medium in electronic equipment for military use, have also occurred in aircraft and submarine guidance systems. Another common problem is corrosion damage suffered by components music players, when exposed to humid environments contaminated with chlorides, for example, during transport by ship, from the manufacture location to the consumer place. Thin layers of corrosion products on the surface of the metal component change their electrical characteristics: resistance, capacity and lead to partial or total failure of the electronic system. There are reported cases where small amounts of moisture have caused corrosion in tablets with printed circuits, nichrome resistors, fittings, electrical connectors and a wide range of components, and micro-electronic components, which have been coated with metallic films (Valdez B. et al., 2006, G.Lopez et. al., 2007)

Corrosion of metal components in the electronics industry may occur at different stages: during manufacture, storage, shipping and service. The main factors in the onset of corrosion and subsequent development are moisture and corrosive pollutants, such as chlorides, fluorides, sulfides and nitrogen compounds, organic solvent vapors, emanating from the resins used as label, or coatings formed during the curing process and packaging of microcircuits.

The sources providing aggressive pollutants are diverse, from flux residues used for welding processes, waste and vapors from electrolytic baths, arising volatile organic adhesives, plastics and acidification of their environment. Assays in artificial atmosphere, which simulates an indoor environment of an electronic plant have shown that the surface of the silver undergoes browning or tarnishing and the formation of dendrite whiskers due to corrosion (Figure 3).

The elemental chemical analysis of the surface (EDX - Scattered Electron Spectroscopy and XRD - X-rays) shows that the corrosion product formed on the silver surface is silver sulfide (Ag_2S), due to the action of pollutant gases such as SO_2 and H_2S present in a humid environment (Figure 4). Moreover, the micrograph of the silver surface (SEM) shows a dendritic growth of corrosion products, characteristic for silver components.

The design of electronics equipment requires a great variety of different metals, due to their different physical and electrical features. Metals and alloys used in the electronics industry are:

- Gold (Au) coating and / or foil in electrical connectors, printed circuits, hybrid and miniature circuits.;
- Silver (Ag) for protective coating in contact relays, cables, EMI gaskets, etc..;
- Magnesium (Mg) alloys for radar antenna dishes and light structures, chassis brackets, etc.;
- Iron (Fe), steel and ferroalloys for guide components, magnetic shielding, magnetic coatings memory disks, processors, certain structures, etc..;
- Aluminum (Al) alloys for armor equipment, chassis, mounting frames, brackets, trusses, etc.;
- Copper and its alloys for cables, tablets printed circuit terminals, nuts and bolts, RF packaging, etc..;
- Cadmium (Cd) for sacrificial protective coating on iron and safe electrical connectors;
- Nickel (Ni) coating for layers such as barrier between copper and gold electrical contacts, corrosion protection, electromagnetic interference applications and compatibility of dissimilar material joints;

- Tin (Sn) coating for corrosion protection of welding; for compatibility between dissimilar metals, electrical connectors, RF shielding, filters, automatic switching mechanisms;
- Welding and weld coatings for binding, weldability, and corrosion protection.

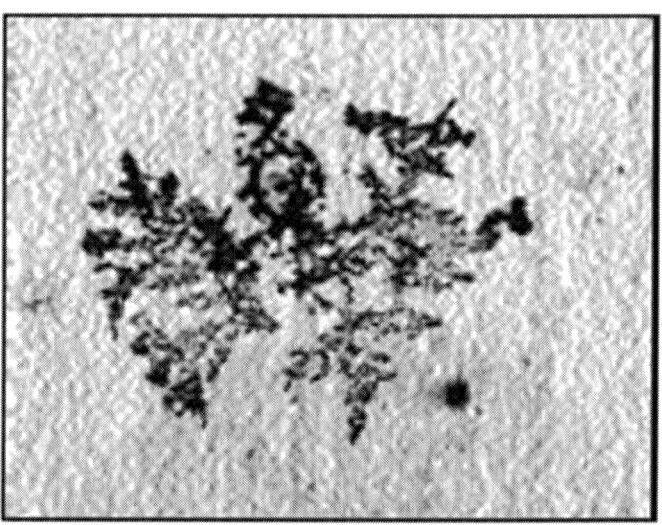

Figure 3: Silver sulfide whiskers corrosion products on silver exposed in an electronics plant atmosphere.

Many of these metals are in contact with each other, so that in the presence of moisture, galvanic corrosion / bimetallic corrosion occurs. When using similar metals, due to design the following requirements must be taken into account.

- Designing the contact of different metals such that the area of the more noble cathodic metal should be appreciably smaller than the area of the more active anodic metal. The area of the cathode can be decreased by applying paint or coating.
- Coating the contact area of a metal with a compatible metal.
- Interpose between dissimilar metals in a metal compatible packaging.
- Sealing interfaces to prevent ingress of moisture.
- Set the electronic device in a hermetically sealed arrangement.

Other corrosion problems can occur due to the characteristics of electronic components such as electromagnetic interference, electromagnetic pulse, flux residues, finishes and materials component tips, organic products that are used for various purposes and emitting gases during curing, whiskers, embrittlement inter-metallic electrical contacts.

Metal components may corrode during manufacture and storage prior to assembly, needing protection against corrosion. In plants and warehouses, air conditioning systems must operate efficiently, removing moisture and suspended particulate matter. Filters and traps should be cleaned and replaced regularly. For closed containers, we recommend the installation of dryers with visual indicators, and the use of volatile vapor phase corrosion inhibitors. In the case of sealed black boxes, the temperature inside these drops should never be below the dew point (Veleva L. et al., 2008, Vargas L. et al., 2009, Lopez G. et al., 2010).

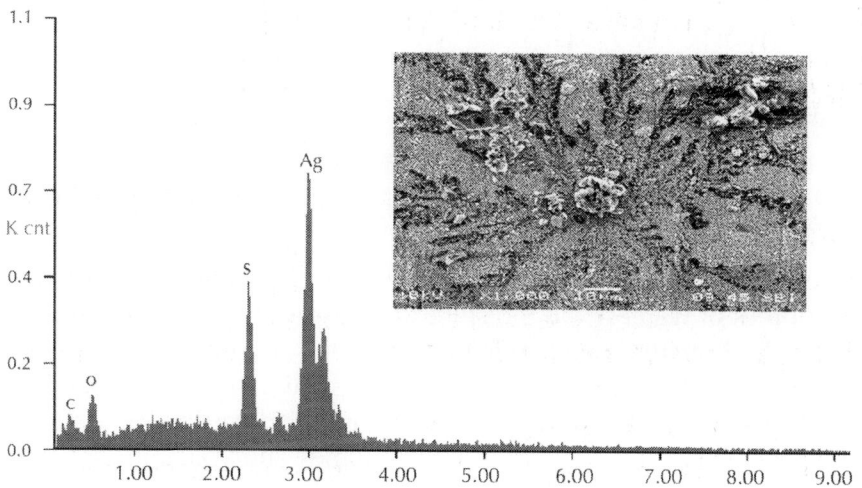

Figure 4: Scanning Electron Microscopy and EDS analysis of silver corrosion products at indoor conditions of an electronics plant contaminated with H_2S.

CORROSION IN WATER

Abundant water sources are essential to a country's industrial development. Large quantities of this precious liquid are required for cooling products, machinery and equipment, to feed boilers, meet health needs and provide drinking water to humans. Estimates of water consumption for each country are different and depend on the degree of industrial development thereof. In first world countries like the United States, these intakes are as high as several hundred billion liters per day. These countries have implemented water reuse systems with certain efficiency due to the application of appropriate treatment for purification. Water, a natural electrolyte is an aggressive environment for many metals / alloys, so that they may suffer from corrosion, whose nature is electrochemical.

As raw water or fresh water we mean natural water from direct sources such as rivers, lakes, wells or springs. Water has several unique properties and one of these is its ability to dissolve to some degree the substances found in the earth's crust and atmosphere allowing the water to contain a certain amount of impurities, which causes problems of scale deposition on the metal surface, e.g. in pipelines, boiler tubes and all kinds of surfaces that are in contact with water (Valdez, B. et al., 1999, 2010).

Oxygen is the main gas dissolved in water, it is also responsible for the costly replacement of piping and equipment due to its corrosive attack on metals in contact with dissolved oxygen (DO). The origin of all sources of water is the moisture that has evaporated from the land masses and oceans, then precipitated from the atmosphere. Depending on weather conditions, water may fall as rain, snow, dew, or hail. Falling water comes into contact with gases and particulate matter in the form of dust, smoke and industrial fumes and volcanic emissions present in the atmosphere.

The concentrations of several substances in water in dissolved, colloidal or suspended form are low but vary considerably. A water hardness value greater than 400 parts per million (ppm) of calcium carbonate, for example, is sometimes tolerated in the public supply, but 1 ppm of dissolved iron should be unacceptable. In treated water for high pressure boiler or where radiation effects are important, as in nuclear reactors, impurities are measured in very small amounts such as parts per billion (ppb).

In the case of drinking water the main concern are detailed physicochemical analysis, to find contamination, and biological assays to detect bacterial load. For industrial water supplies it is of interest the analysis of minerals in particular salts. The main constituents of water are classified as follows:

- Dissolved gases: oxygen, nitrogen, carbon dioxide, ammonia and sulfide gases;
- Minerals: calcium, sodium (chloride, sulfate, nitrate, bicarbonate, etc.), Salts of heavy metals and silica;
- Organic matter: plant and animal matter, oil, agricultural waste, household and synthetic detergents;
- Microbiological organisms: include various types of algae, slime forming bacteria and fungi.

The pH of natural waters typically lies within the range of 4.5 to 8.5; at higher pH values, there is the possibility that the corrosion of steel can be suppressed by the metal passivation. For example, Cu is greatly affected by the pH value in acidic water and undergoes a slight corrosion in water releasing small amounts of Cu in the form of ions, so that it's corroded surface because green stained clothing and sanitary ware. Moreover, deposition of the Cu ions on surfaces of aluminum or galvanized zinc corrosion cells leads to new bimetallic contact, which cause severe corrosion in metals.

The mineral water saturation produces a greater possibility of fouling on the metal walls, due to the ease with which the insoluble salts (carbonates) can be precipitated. To control this effect it is necessary to know and use the Saturation Indices. Water saturation refers to the solubility product of a compound and is defined as the ratio of the ion activity and the solubility product. For example, water is saturated with calcium carbonate when it is no more possible to dissolve the salt in water and then it begins to precipitate as scale. In fact, it is called supersaturated when carbonate precipitation occurs on standing the solution. The most common parameters that must be known to characterize the water corrosivity, be it raw or treated, for operation in an industrial facility are shown in Table 2.

Table 2: Water properties and corrosivity

Water properties	Corrosivity
Hardness	Source of scaling that promotes corrosion
Alkalinity	Produces foam and motion of solids
pH	Corrosion depends on its value
Sulphates	Produces scaling
Chloride	Increases water corrosivity
Silica	Generates scaling in hot water. Condensers and steam turbines
Total Dissolved Solids (TDS)	Increases electrical conductivity and corrosivity
Temperature	Elevated temperatures increases corrosion rates

There six formulas to calculate Saturation Indices and embedding: Langelier index (LSI), Ryznar stability index, Puckorious index of scaling, Larson-Shold index, index of Stiff- Davis and Oddo-Tomson index. There is some controversy and concern for the correlation of these indices with the corrosivity of the waters, particularly regarding the Langelier (LSI).

A LSI saturation index with value "0" indicates that the water is balanced and will not be fouling, while the positive value indicates that the water may be fouling (Table 3). The negative value of the LSI suggests that water is corrosive and can damage the metal installation, increasing the content of metallic ions in water. While some sectors of the water management industry uses the values of the indices as a measure of the corrosivity of the water. Corrosion specialists are alerted and are very wary of issuing an opinion, or extrapolate the use of indices to measure the corrosivity of the environment.

Table 3: Langelier index for water corrosivity and scaling

Langelier Index	Water corrosivity and scaling
-5.0	Severe corrosion
-4.0 to -2.0	Moderate corrosion
-1.0 to -0.5	Light corrosion
0.0	No corrosion / no scale (balance)
0.5 to 2.0	Light incrustation
3.0	Moderate incrustation
4.0 to 5.0	Severe incrustation

Sometimes the raw water is contaminated with chemicals such as fertilizers and other chemicals coming from agricultural areas (Figure 5).

In these cases, ionic agents such as nitrites, nitrates, etc., in water causes an accelerated process of localized corrosion to many metals and the consequent failure of equipment.

Figure 5: Corrosion on the gates dam on the Colorado River, Baja California, Mexico.

Raw water contaminants can be quite varied, including both heavy metals and organic chemicals, referred to as toxic pollutants. Among the heavy metals may be mentioned arsenic (As), mercury (Hg), cadmium (Cd), lead (Pb), zinc (Zn) and cadmium (Cd), which are sometimes at trace levels, but they tend to accumulate over time, so that priority pollutants are to be treated.

Pesticides, insecticides and plaguicides comprise a long list of compounds, for which we should be concerned: DDT (insecticide), aldrin (an insecticide), chlordane (pesticide), endosulfan (insecticide), diazinon (insecticide), among others.

Contaminants, such as polycyclic aromatic organic compounds, include what is known as volatile organic compounds such as naphthalene, anthracene and benzopyrene. There are two main sources of these pollutants: petroleum and combustion products found in municipal effluents. On the other hand, there are polychlorinated biphenyls or PCBs, which are mainly used in transformers for the electrical industry, heavy machinery and hydraulic equipment. This class of chemicals is extremely persistent in the environment and affects human health.

From the viewpoint of corrosion, these contaminants which are present even at low concentrations or trace in the raw water, favor the corrosivity the metals which are in contact with. The combination of the corrosive effects of these contaminants together with the oxidation by oxygen, minerals and other impurities, leads to consider raw water as a natural means capable of generating corrosion of metals. It is recommended at least, to carry out a process of treating raw water, to reduce significantly the hardness and remove suspended solids, which will help greatly in preventing subsequent problems of corrosion and fouling on metal surfaces, curbing economic losses and maintaining the industrial process in good operating condition.

Corrosion in Potable Water Systems

Corrosion is a complex phenomenon that arises as a result of the interaction between water and the surface of metallic pipes or the equipment of storage and handling. The process is invariably a combination of oxidation and reduction, as already described in previous chapters. In drinking water, it should be noted that the corrosion products which are partially soluble in water in ionic form are toxic at certain concentrations, e.g. copper and lead. The existence of high concentrations of lead in water carried by copper tubing, indicate that the source of lead may be tin-lead solder at the junctions of the copper pipes. The consumption of domestic water contaminated with toxic metal ions (Pb^{+2}, Cu^{+2}, Zn^{+2}, Cr^{+3}), gives rise to acute chronic health problems. The regulations have set the following limits allowable concentration in drinking water: Cr (0.05 ppm), Cu (0.01 ppm), Pb (0.05 ppm) and Zn (5 ppm). These regulations are made in order to protect the public user and consumer of drinking water and are continuously striving for a reduction in the maximum allowable limits. Some concentrations reach zero as is the case of Pb in the United States due to the concerns Pb about poisoning of children. Still, many sources such as wells and springs are outside the control of law and toxic substances, bacteria and pathogens. Damage caused by corrosion of household plumbing may be accompanied by unpleasant aesthetic problems such as soiled clothing, unpleasant taste, stains and deposits in the toilets, floors of bathrooms, tubs and showers. To prevent corrosion of pipes, we recommend the use of PVC pipes for drinking water, replacing the metal, as a preventive measure.

Corrosion can occur anywhere on the pipes that carry drinking water, mainly at sites of contact between two dissimilar metals, thus forming a corrosion cell. In general, the metals will corrode to a greater or lesser degree in water, depending on the nature of the metal, on the ionic composition of water and its pH. Waters high in dissolved salts (water hardness), favor the formation of scale, more or less adherent, in different parts of the equipment (Figure 6). These deposits may be hard or brittle, sometimes acting as cement, creating a physical barrier between the metal and water, thereby inhibiting corrosion. Calcium carbonate ($CaCO_3$) is the most common scale; its origin is associated with the presence of carbon dioxide gas (CO_2) in water. Sometimes these deposits are filled with pasty or gelatinous hydrated iron oxides or colonies of bacteria (Valdez, B. et al., 1999, 2010).

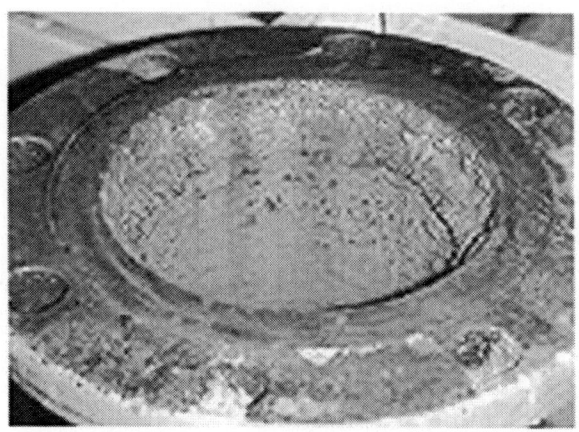

Figure 6: Corrosion in potable water pipes.

Usually, groundwater $CaCO_3$ saturated (calcareous soils), due to the presence of dissolved CO_2, whose content depends on its content in the air in contact with the water and on temperature. These waters are often much higher in CO_2 content, so they may dissolve substantial amounts of calcium carbonate. These waters are at pressures lower than they had in the ground, so CO_2 gas lost with consequent supersaturation of carbonates. If conditions are appropriate, the excess of $CaCO_3$ can precipitate as small agglomerates deposited in muddy or hard layers on solid surfaces, forming deposits. An increase in temperature is an important factor and also leads to supersaturation of carbonates, with the consequent possibility of fouling. To a lesser extent fouling can precipitate more soluble Mg carbonates ($MgCO_3$) and Mn ($MnCO_3$), and also oxides / hydroxides, dark colored and gelatinous. Except in very exceptional cases in sulfated water, it is normal to find deposits of gypsum ($CaSO_4 \bullet \frac{1}{2} H_2O$) because their solubility is high, but decreases with increasing temperature. Hard silica scale (SiO_2) may appear with oversaturated waters or appear as different silicates (SiO_4^{4-}) trapped in the carbonate deposits. Generally, the silica appears trapped in other types of scale and it is not chemical precipitation.

Waters often carry considerable amounts of iron (ferrous ion, Fe^{+2}), which may be often precipitated by oxidation upon contact with air as hydrated iron oxide (ferric, Fe^{+3}) but sometimes can be Fe^{+2} form black sludge, more or less pasty or gelatinous and sometimes very large. The voluminous precipitate occupies the pores, significantly reducing the permeability of the fouling. Sometimes the Fe ions can come from corrosion of the pipe giving rise to simultaneous corrosion and scaling (Figure 6). Common bacteria of the genera Gallionella, Leptothrix Cremothrix are known as Fe bacteria, can give reddish-yellow voluminous precipitate and sticky ferric compounds from ferrous ion, which drastically reduce the permeability of the deposit, in addition to trap other insoluble particles.

The cost for impairment of domestic water systems and the impact on health, involves several consequences: premature corrosion and failure of the pipes and fittings that carry water in a house or building, a low thermal efficiency (up to 70%) of water heaters (boilers), which can cause their premature failure. High levels of metals or oxides, which usually are not properly, treated in drinking water cause red or blue-green deposits and stains in the toilets sinks. In addition to concerns about the aesthetic appearance, a corrosion process can result in the presence of toxic metals in our drinking water. For evaluating water quality and their tendency corrosive and / or fouling, LSI can be used. This analysis must be accompanied by measurements of water pH and conductivity, and corrosion tests applying international standards.

Anticorrosive Treatment of Water

Corrosion control is complex and requires a basic knowledge of corrosion of the system and water chemistry. Systems can be installed for water pretreatment, using non-conductive connections, reducing the temperature of hot Cu water pipes employed and copper installing PVC or other plastic materials. It is important to note that the corrosiveness of water can be increased by the use of water softeners, aeration mechanisms, increasing the temperature of hot water, water chlorination, and attachment of various metals in the water conduction system. A proper balance between the treatment systems and water quality, can be obtained with acceptable levels of corrosivity. Thus, the lifetime of the materials that make the water system in buildings, public networks, homes and other systems will be longer.

SOIL CORROSION

A large part of steel structures: aqueducts, pipelines, oil pipelines, communications wire ropes, fuel storage tanks, water pipes, containers of toxic waste, are buried, in aggressive soils. Large amounts of steel reinforced concrete structures are also buried in various soil types. In the presence of soil moisture it is possible to have humid layer on the metal surface, whose aggressiveness depends on soil type and degree of pollution (decaying organic matter, bacterial flora, etc.). Thus, the soil can form on the metal surface an electrolyte complex with varying degrees of aggressiveness, a necessary element for the development of an underground electrochemical corrosion. The corrosion process of buried structures is extremely variable and can occur in a very fast, but insignificant rate, so that pipes in the soil can have perforations, presenting localized corrosion attack or uniform.

Metal structures are buried depending on their functionality and security. Most often they traverse large tracts of land, being exposed to soils with different degrees of aggressiveness exposed to air under atmospheric conditions (Figure 7).

Figure 7: Valve system of a desert water aqueduct.

When pipes or tanks are damaged by corrosion, the formation of macro-and micro-cracks can lead to leaks of contained products or fluids transported, causing problems of environmental pollution, accidents and explosions, which can end in loss of life and property (Guadalajara, Jalisco, Mexico, 1992). In the case of pipes used to carry and distribute water, a leak may cause loss of this vital liquid, so necessary for the development of society in general and especially important in regions where water is scarce, so the leakage through aqueducts pipes should be avoided. An important tool needed to prevent the most serious events, is the knowledge of the specific soil and its influence on the corrosion of metal structures.

Types of Soils and Their Mineralogy

A natural soil contains various components, such as sand, clay, silt, peat and also organic matter and organisms, gas, mineral particles and moisture. The soils are usually named and classified according to the predominant size range of individual inorganic constituent particles. For example, sandy soil particles (0.02 - 2 mm) are classified as fine sand (0.02 - 0.2 mm) or thick (0.20 -2.00 mm). Silt particles (0.002 to 0.02 mm) and clay, which have an average diameter 0.002 mm, are classified as colloidal matter. A comparison of the sizes of these typical soils is done in Figure 8.

Currently exists in the U.S. and in over 50 countries worldwide, a detailed classification for soils, which includes nine classes with 47 subgroups.

The variation in the proportion of the groups of soil with different sizes, determines many of its properties. Fine-textured soils due to high clay content, have amassed particles, so they have less ability to store and transport gases such as oxygen, that any ground-open e.g. sandy soil. The mineralogy of both clay types and their properties, are closely related to the corrosivity of the soil. Silica (SiO_2) is the main chemical constituent of soils type clay, loam and silt, also in the presence of Al_2O_3. Common species in moist soil are dissolved ions H^+, Cl^-, SO_4^{2-}, HCO_3^-. The chemical composition and mineralogy of the soil determine its corrosive aggressiveness; poorly drained soils (clay, silt and loam) are the most corrosive, while soils with good drainage (gravel and sand type) are less aggressive to metals. Vertically homogeneous soils do not exist, so it is convenient to consider the non-uniformity of ground, formed of different earth layers. To understand the corrosion behavior of a buried metal is very important to have information about the soil profile (cross section of soil layers). The physicochemical and biological nature of soil, corrosive aggressiveness and dynamic interactions with the environment, distinguishes the ground like a very complex environment and different from many others. Climate changes of solar radiation, air temperature and relative humidity, amount of rainfall and soil moisture are important factors in corrosion. Wind, mechanical action of natural forces, chemical and biological factors, human manipulation can alter soil properties, which directly affects the rate of corrosion of metals buried in the ground. Conditions may vary from atmospheric corrosion, complete immersion of the metal, depending on the degree of compactness of the soil (existence of capillaries and pores) and moisture content. Thus the variation in soil composition and structure can create different corrosion environments, resulting in different behavior of the metal and oxygen concentrations at the metal / soil interface.

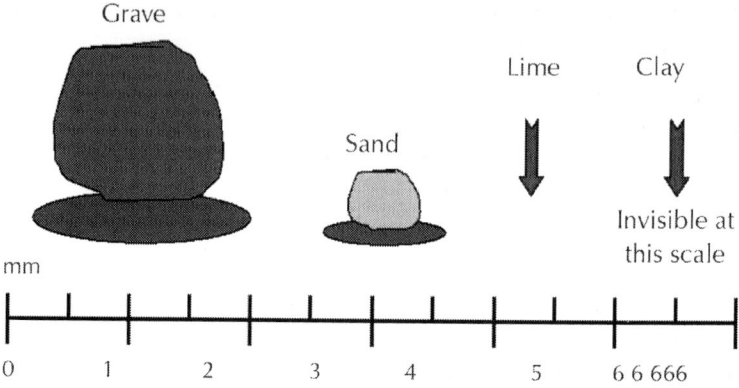

Figure 8: Size of soil particles.

Two conditions are necessary to initiate corrosion of metal in soil: water (moisture, ionic conductor) and oxygen content. After startup, a variety of variables can affect the corrosion process, mentioned above, and among them of importance are the relative acidity or alkalinity of the soil (pH), also the content and type of dissolved salts.

Mainly three types of water provide moisture to the soil: groundwater (from several meters to hundreds below the surface), gravitational (rain, snow, flood and irrigation) and capillary (detained in the pores and capillary spaces in the soil particles type clay and silt). The moisture content in soils can be determined according to the methodology of ASTM D 2216 ("Method for Laboratory Determination of Water (Moisture) Content of Soil and Rock by Mass"), while its permeability and moisture retention can be measured the methods described in ASTM D2434 and D2980. The presence of moisture in soils with a good conductivity (presence of dissolved salts), is an indication for high ion content and possible strong corrosive attack.

The main factors that determine the corrosive aggressiveness of the soil are moisture, relative acidity (pH), ionic composition, electrical resistance, microbiological activity.

Corrosion Control of Buried

Given the electrochemical nature of corrosion of buried metals and specific soils, this can be controlled through the application of electrochemical techniques of control, such as cathodic protection. This method has been universally adopted and is appropriate to protect buried metallic structures. For an effective system of protection and cheaper maintenance, pipelines must be pre-coated, using different types of coatings, such as coal tar, epoxies, etc. This helps reduce the area of bare metal in direct contact with the ground, lowering the demand for protection during the corrosion process. The purpose of indirect inspection is to identify the locations of faulty coatings, cathodic protection and electrical Insufficient shorts (close-interval, on/off Potential surveys, electromagnetic surveys of attenuation current, alternating current voltage gradient surveys, etc..), interference current, geological surveys, and other anomalies along the pipeline.

CORROSION UNDER THERMAL INSULATION

One of the most common corrosion problems in pipes, ducts, tanks, preheaters, boilers and other metal structures, insulated heat exchange systems, is the wear and corrosion occurring on metal (steel, galvanized steel, Al, SS, etc.), below a deposit or in its immediate neighborhood. This corrosion is known as corrosion under deposit. The deposit may be formed by metal corrosion products and / or different types of coating applied for protection. For example, in the case of a calcareous deposit, formed in the walls of galvanized steel pipes which carry water with a high degree of hardness (dissolved salts), it might develop corrosion under deposit. These shells may be porous, calcareous deposit and / or partially detached from the metal surface, so that direct contact between metal, water and oxygen (the oxidizing agent in the corrosion process) allows the development of metal corrosion. For this reason the pipes could be damaged severely in these locations up to perforation, while in parts of the installation corrosion might occur at a much lower level.

There is a considerable amount of factors in the design, construction and maintenance, which can be controlled to avoid the effects of deterioration of metal by corrosion under deposit. In general, under these conditions the metal is exposed to frequent cycles of moisture, corrosivity of the aqueous medium or failure in the protective coatings (paint, metal, cement, fiberglass, etc.). Figure 9 shows a conductor tube steam in a geothermal power plant, where CS corrosion happened beneath the insulation.

Figure 9: Corrosion of a carbon steel pipe under insulation.

Seven factors can be controlled on the ground, to prevent this type of corrosion: design of equipment, operating temperature, selection of the insulation, protective coatings and paints, physical barriers from the elements, climate and maintenance practices of the facility. Any change in any of these factors may provide the necessary conditions for the corrosion process to take place. The management knowledge of these factors help explain the causes of the onset conditions of corrosion under deposits, and it will guide a better inspection of existing equipment and the best design.

Equipment Design

The design of pressure vessels, tanks and pipes, generally includes accessories for support, reinforcement and connection to other equipment. Details about the installation of accessories are the responsibility of the engineers or designers, using building codes to ensure reliability of both insulated and non insulated equipment. The protective barrier against the environment surrounding the metal structure in such designs often breaks donor due to an inappropriate insulation, loss of space for the specified thickness of insulation or simply by improper handling during installation of the equipment. The consequence of a rupture or insulation failure means greater flow water ingress to the space between metal and coating hot-cold cycle, generating over time a buildup of corrosive fluid, increasing the likelihood of corrosive damage. Moreover, wet insulation will be inefficient and also cause economic losses. The solution of this factor is to meet the thickness specifications and spacing, as indicated in the code or equipment-building specifications and characteristics of the coating used.

The operating temperature is important for two reasons: a high temperature favors the water is in contact with the metal for less time, however, also provides a more corrosive environment, causes fast failures of coatings. Usually a team operating in freezing temperatures is protected against corrosion for a considerable life time. However, some peripheral devices, which are coupled to these cold spots and operating at higher temperatures, are exposed to moist, air and steam, with cycles of condensation in localized areas, which make them more vulnerable to corrosion. For most operating equipment at freezing conditions, the corrosion occurs in areas outside and below the insulation. The temperature range where this type of corrosion occurs is 60 °C to 80 °C; however, there have been failures in zones at temperatures up to 370 °C. Also, in good water-proof insulation, corrosion is likely to occur at points where small cracks or flaws are present, so that water can reach the hot metal and evaporate quickly. On the other hand, in machines where the temperature reaches extreme values, as in the case of distillation towers, it is very likely to occur severe corrosion problems.

Selection of Insulation

The characteristics of the insulation, which have a greater influence on the corrosion processes deposits, are the ability to absorb water and chemical contribution to the aqueous phase. The polyurethane foam insulation is one of the most widely used; however, in cold conditions they promote corrosion due to water absorption present. The coatings of glass fiber or asbestos can be used in these conditions, always when the capacity of absorbing water do not becomes too high. Corrosion is possible under all these types of coating, such insulation. The selection of insulation requires considering a large group of advantages and disadvantages regarding the installation, operation, cost, and corrosion protection, which is not an easy task. The outside of the insulation is the first protective barrier against the elements and this makes it a critical factor, plus it is the only part of the system that can be readily inspected and repaired by a relatively inexpensive process. The durability and appearance, melting point fire protection, flame resistance and installation costs are other important factors that must be taken into account together with the permeability of the insulation. Usually the maintenance program should include repairs to the range of 2 to 5 years. Obviously the weather is important and corrosion under thermal insulation will more easily in areas where humidity is high. Sometimes conditions of microclimate can be achieved through the use of a good design team.

CORROSION IN THE AUTOMOTIVE INDUSTRY

One of the most important elements of our daily life, which has great impact on economic activity, is represented by automotive vehicles. These vehicles are used to transport people, animals, grains, food, machinery, medicines, supplies, materials, etc. They range from compact cars to light trucks, heavy duty, large capacity and size. All operate mostly through the operation of internal combustion engines, which exploit the heat energy generated by this process and convert it in a mechanical force and provide traction to these vehicles.

The amount and type of materials used in the construction of automotive vehicles are diverse, as the component parts. They are usually constructed of carbon steel, fiberglass, aluminum, magnesium, copper, cast iron, glass, various polymers and metal alloys. Also, for aesthetic and protection against corrosion due to environmental factors, most of the body is covered with paint systems, but different metal parts are protected with metallic or inorganic coatings.

Corrosion in a car is a phenomenon with which we are in some way familiar and is perhaps for this reason that we often take precautions to avoid this deterioration problem.

A small family car, with an average weight of 1000 kg, is constructed of about 360 kg of sheet steel, forged steel 250 kg, 140 kg cast iron mainly for the engine block (now many are made of aluminum), 15 kg of copper wires, 35 kg and of plastic 50 kg of glass that usually do not deteriorate, and 60 kg for rubber tires; which wear and tear. The remaining material is for carpets, water and oil. Obviously, that is an advanced technology in the car industry, with automobiles incorporating many non-metallic materials into their structure. However, the problem of corrosion occurs at parts where the operation of the vehicle is compromised. Corrosion happens in many parts of the car (mostly invisible) it is not only undesirable for the problems it causes, but also reduces the vehicle's resale value and decreases the strength of the structure. To keep the car in good condition and appearance, its high price, it is necessary to pay attention to the hidden parts of the vehicle.

The main cause of corrosion of the car body is the accumulation of dust in different closed parts, which stays for a long time by absorbing moisture, so that in these areas metal corrosion proceeds, while in the clean and dry external parts it does not occur (Figure 10).

The corrosion problem that occurs in the metal car body has been a serious problem that usually arises most often in coastal environments, contaminated with chlorides and rural areas with high humidity and specific contaminants. Many countries use salt ($NaCl$, $CaCl_2$ or $MgCl_2$) to keep the roads free of ice; under these conditions these salts, in combination with the dust blown by the car, provide conditions for accelerated corrosion. Therefore, it is recommended as a preventative measure, after a visit on the coast or being on dirty roads, to wash the car with water, and also the tires and the doors, especially their lower parts. In urban environments, the corrosion problem has been reduced due to the new design and application of protective coatings, introduced by major manufacturers in the early nineties of the twentieth century. The areas most affected are fenders, metal and chrome bumpers views which are used in some luxury vehicles as well as areas where water and mud are easily accumulated e.g. auctions of funds windshield and doors (Figure 11).

In regions with high incidence of solar radiation and the presence of abrasive dust, paint vehicles deteriorate rapidly. The hot, humid weather, combined with high levels of SO_2 and NO_x emissions that come from burning oil, chlorides salt. In the Gulf of Arabia, the blowing sand from the nearby desert, creates a very aggressive environment; statistics reveals that one in seven cars is damaged and due to corrosion the car life is estimated to an average of 8 months, also the car corrosion resistance decreases in the following order: manufactured in Europe, USA and Japan. White paints generally have shown a significantly better corrosion protection than other colors. Initially, corrosion defects appear as a kind of dots and spots of corrosion products formed under the paint and subsequently emerge from the steel sheet, leaving a free entry for moisture and air (oxygen), accelerating the corrosion process; in these cases reddish metal corrosion products.

Figure 10: Corrosion on a bodywork exposed to the Gulf of Mexico tropical coast.

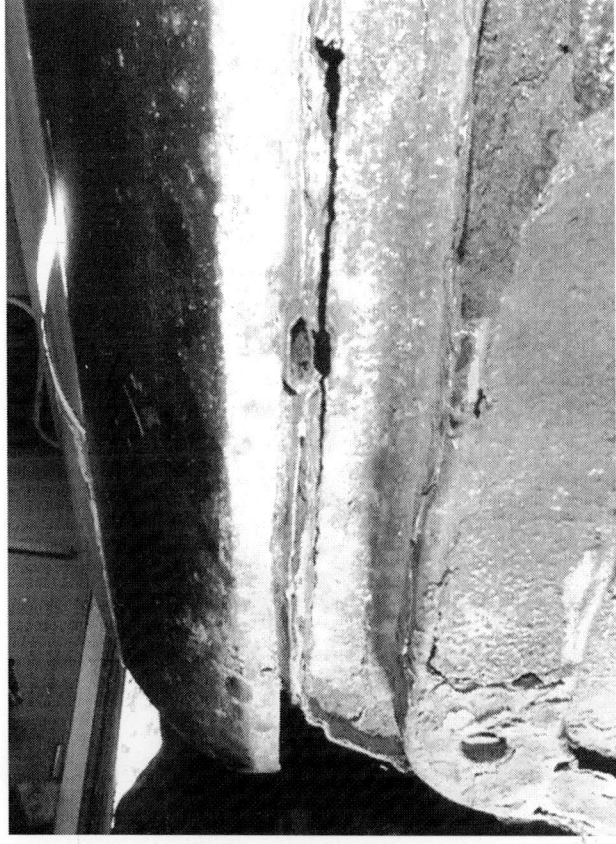

Figure 11: Corrosion on a car door and bottom of the bodywork.

Corrosion in the Cooling System

The cooling system of a car combustion engine consists of several components, constructed of a variety of metals: radiators are made of copper or aluminum, bronze and solder couplings with tin water pumps; motors are made of steel, cast iron or aluminum. Most modern automobiles, with iron block engine and aluminum cylinder head, require inhibitor introduced into the cooling water to prevent corrosion in the cooling system. The inhibitor is not antifreeze, although there are in the market solutions which have the combination of inhibitor-antifreeze. The important thing is to use only the inhibitor recommended in the automobile manual and not a mixture of inhibitors, since these may act in different ways and mechanisms. The circulating water flow should work fine without loss outside the system. If the system is dirty, the water should be drain and filling the system with a cleaning solution. It is not recommended to fill the system with hard water, but with soft water, introducing again the inhibitor in the correct concentration. If there exhaust at the water cooling system, every time water is added the inhibitor concentration should be maintained to prevent.

In small cars, it is common for water pumps; constructed mainly of aluminum, to fail due to corrosion, cavitation, erosion and corrosion, making it necessary to replace the pump (Valdez, B. et al., 1995). Accelerated corrosion in these cases is often due to the use of a strong alkaline solution of antifreeze. On the other hand, in heavy duty diesel trucks, the cooling system is filled with tap water or use filters with rich conditioner chromates that can cause the pistons jackets to suffer localized corrosion. After 12 or 15 months, the steel jackets are perforated and the water passes into the cavity through which the piston runs, forcing to carry out repair operations (Figure 12).

Figure 12: Corrosion in a carbon steel jacket on the water face in a diesel combustion engine truck.

Corrosion causes great economic losses to the transport industry, since it must stop to repair the truck and abandon to provide the service with all the consequences that this entails. Furthermore, the use of chemical conditioning is now controlled by environmental regulations, so chromates and phosphates are restricted and novel mixtures of corrosion inhibitors have been produced to control the problem of corrosion in automobile cooling systems.

Corrosion in Exhaust Pipes and Batteries

Exhaust pipes made of SS (0.6 - 0.8 mm thick) have a better resistance to chemical corrosion at high temperatures, which is why we are now using SS in many popular models. This SS resists corrosion much more than conventional CS and thus their long life covers the higher price. Another alternative is to use conventional CS tube, zinc coated or aluminum (Figure 13). These exhaust pipes are less expensive than stainless steel, but less resistant to corrosion.

Figure 13: Corrosion on carbon steel exhaust pipes coated with aluminum.

The acidic environment which is generated on the surface of accumulators supplying the energy necessary for starting the engine, favors conducting corrosion processes in the lead terminals, where the cables are connected by bronze or steel clamps. Thus, this environment and these contact zones predispose cells to a process galvanic corrosion, which gradually deteriorates the contact wires, generating bulky corrosion products. This phenomenon is called sulfation of the contacts due to the sulfuric acid containing the battery, thus forming white sulfates on the corroded metal surface. These products introduce high resistance to current flow and cause failure to the engine ignition system, and impede the battery charge process. This problem has been eliminated in batteries that have airtight seals, or are manufactured with new technologies as well as bases covered with organic coatings that prevent corrosion.

Some years ago it was common for starters to fail, because the moisture or water penetrated into the gear area preventing it sliding motion and causing burning of the electric motor. Currently, new designs avoid contact with moisture and other foreign agents, preventing the occurrence of corrosion problems in these devices. As a preventive measure is recommended to prevent spillage of battery acid, to periodically clean the battery terminals (with a brush of wire or a special instrument), also coat them with petroleum jelly to prevent corrosion in these contact areas. A fat based composition which contains several components: alkaline salts and oxides of lithium, sodium bicarbonate and magnesium oxide are applied to the terminals and the connector. In general, in wet weather, the contacts of the accumulators have a tendency to more accelerated corrosion, thus requiring greater care to disconnect the terminals when not being used.

Corrosion Prevention

To keep the vehicle for a longer time without the appearance of corrosion, it always requires washing with running water and, the use of very soft brush or cloth-like material, with a special detergent (not household detergents, which are very corrosive) and finally wash the vehicle with plenty of water. The floor carpet should be maintained clean and dry. A car should not be left wet in a hot garage, since under these conditions accelerated corrosion takes place since the water does not dry and can condense on the cold parts of the vehicle. In these cases, it is best not to close the garage door or use a roof space, to protect it from rain, and not allow moisture condensation. However, if the vehicle is left unused for a long time in a closed garage, it should be protected from dust, moisture and contaminants.

CORROSION CONTROL IN THERMOELECTRIC PLANTS

Electricity is a key element in ensuring economic growth and social development of a country. Many conventional power plants in recent years are being installed in combined cycle power plants, also called cogeneration. The latter, simultaneously generate electricity and / or mechanical power and useful heat, sometimes using thermal energy sources that are lost in conventional plants.

A power station is a thermoelectric energy conversion system, starting with the chemical energy of fuel that during combustion is converted into heat energy accumulated in the steam. This thermal energy generates mechanical energy from the hot steam, which expands in a turbine, turning on electricity in the generator. In this process of low energy thermal efficiency is lost in the hot gases that escape through the chimney and the cooling steam in the condenser.

Electricity generating plants burn fossil fuels such as coal, fuel oil and natural gas. These fuels containing as minor components sulfur compounds (S), nitrogen (N), vanadium (V) and chloride (Cl^-). These are corrosive chemicals attacking the metal infrastructure; and polluting the environment by becoming acid gas emissions, also affecting the health of the population.

The three central equipment of a thermoelectric plant are the boiler, which converts the water into steam, the steam turbine to whom the pressure imparts a rotary motion and the condenser that condenses the vapor released by the turbine and the condensed water is returned to the boiler as feed water. The turbine itself transmits rotary motion to the generator of electricity, which will be distributed to industrial, commercial and homes in cities.

Corrosion in steam plant equipment occurs in two parts of the boiler: on the water side and the steam side, with the fire temperature up to 700 ° C, depending on the type, size and capacity of the boiler. The boiler feedwater must be treated to eliminate the corrosive components: salts such as chlorides and sulfates dissolved oxygen (DO); silicates and carbonates, producing calcareous scale on the boiler walls, regarded as precursors for the formation of corrosion under deposits. The water is softened by eliminating salts and treated to remove oxygen; the pH is controlled by addition of alkaline phosphate to reach a pH range of 10 to 11, and inhibitors are added to the feedwater to prevent corrosion.

The flue gases and ash solid particles reach temperatures up to 1000 to 1200 °C, impinging on the outer surface of the boiler water tubes and preheater, creating an atmosphere for aggressive chemical corrosion. The damaged tubes lose its thickness generating metal corrosion products; they often are fractured, suffering a stress corrosion due to the combined effects of mechanical stress and corrosion (Figure 14). Since the tubes lose steam and pressure, the operation of the plant is interrupted and the tubes or its sections should be changed incurring severe economic losses. For example, in the United States has been concluded that the costs of electricity are more affected by corrosion than any other factor, contributing 10% of the cost of energy produced.

Figure 14: Stress corrosion cracking, in a combustion gases pipe of a thermo-electric station.

A study reveals that in 1991 there were more than 1250 days lost in nuclear plants operating in the United States, due to failure by corrosion, which represented an economic loss of $ 250.000 per day. Such statistics indicate that the power generation industry needs to obtain a balance between cost and methods for controlling effectively corrosion in their plants. It is sometimes advisable to add additives to the fuel, for example, magnesium oxide which prevent the deposition of the molten salts on the boiler tubes. Corrosion occurs also in the combustion air preheater, by sulphurous gases which react with condense and form sulfuric acid. Metal components of the turbine rotor: disks and blades suffer from corrosion by salts, alkalis and solid particles entrained in the vapor. In these cases, it is common to observe the phenomena of erosion-corrosion, pitting and stress corrosion fracture; their damage can be ameliorated through a strict quality control of boiler water and steam.

Efficient maintenance and corrosion control in a power plant is based on the following:

- Operation according to mechanical and thermal regime, indicated by the designer and builder of the plant;
- Correct treatment of fuel, water and steam;
- Chemical cleaning of the surfaces in contact with water and steam, using acidic solutions containing corrosion inhibitors, passivating ammoniacal solutions and solutions;
- Mechanical cleaning of surfaces covered with deposits (deposits), using alkaline solutions and water under pressure;
- Perform an optimum selection of the materials of construction for the components of the plant, including those suitable as protective coatings.
- The installation of online monitoring of corrosion in critical plant areas will be one of the most effective actions to control corrosion. In addition, it is recommended same use and document to use corrosion expert system software and materials databases for the analysis of the materials corrosion behavior.

Corrosion in power plants can be controlled by applying the knowledge, methods, standards and materials, based on corrosion engineering and technology.

CORROSION IN GEOTHERMAL ENVIRONMENTS

The development of alternative energy sources represents one of the most attractive challenges for engineering. There are several types of renewable energies already in operation, such as wind, solar and geothermal. Geothermal environments can lead to aggressive environments, e.g. the geothermal field of "Cerro Prieto", located in Baja California, Mexico.

The physical and chemical properties of the vapor at "Cerro Prieto" make it an aggressive environment for almost any type of material: metal, plastic, wood, fiberglass or concrete. The typical chemical composition of a geothermal brine, is shown in Table 4. Many engineering materials are present as components of the infrastructure and field equipment, required for the steam separation, purification and posterior operations for the generation of electricity. This entire infrastructure is a costly investment and therefore, failure or stoppage of one of them, means economic losses, regardless of how vital it is to maintain constant production of much-needed electricity.

Figure 15: Corrosion in concrete structures used to separate steam from water and to operate steam silencers.

In the process of the geothermal fluid exploitation, corrosion of metal structures occurs from the wells drilling operation, where the drilling mud used, causes corrosion of pumping and piping equipment. Subsequently, when the wells pipes are in contact with the steam, they can also suffer from corrosion-erosion problems, where the corrosive agent is hydrogen sulfide. Steam separators and the pipes are exposed to problems of fouling and localized corrosion due to the presence of aggressive components such as H_2S and chloride ions (Cl^-), present in the wells fluid. These agents lead to the deterioration of reinforced concrete foundations supporting steel pipes, or other concrete structures used to separate steam from water and to operate steam silencers. The reinforced concrete deterioration due to steel corrosion in this aggressive environment, and the steam pressure mechanical forces lead to concrete damage with formation of cracks and fractures.

Table 4: Typical chemical composition of typical "Cerro Prieto" geothermal brine

Component	Na^+	K^+	Mg^{2+}	Ca^{2+}	Cl^-	SO_4^{2-}	Si_o2	HCO_3^-
Ppm, mg/kg	6429	1176	18.6	347	11735	15	1133	303

In the power plants, the observed corrosion affects components of the steam turbines, condensers and pipelines, and also the cooling towers and concrete structures inside and outside the building that houses the plant. In these cases, the effects of corrosive attack appears in the form of localized corrosion in metal walls and gas piping) or as corrosion fatigue or stress corrosion, caused by cyclic mechanical forces or residual stresses, in turbines and other metal equipment. Table 5 shows a list of equipment and materials used for construction, which are part of the infrastructure of a geothermal power (Valdez, B. et al., 1999, 2008)

The combination of an aerated moist environment with the presence of hydrogen sulfide gas (H_2S) dissolved in water provides a very aggressive medium (Figure 16), which promotes the corrosion of metals and alloys, such as CS and SS. The presence of dust, from the geothermal field and condensation cycles favor the failure of protective coatings applied to steel, so that developed corrosion leads to constant repairs and maintenance of metal installations: pipes, machinery, cooling towers, vehicles, tools, fences, warehouses, etc.

Table 5: Equipment and materials used to build infrastructure in a geothermal field

Equipment	Materials
Pipelines	Concrete, steel
Vertical and centrifugal pumps	Steel, copper alloys
Valves	Steel
Flanges and fits	Steel
Silencers	Concrete, steel, FRGP
Brine canals	Reinforced concrete
Evaporation ponds	Plastics
Control and safety instruments	Metals and plastics

Cooling towers constructed of wood, steel and fiberglass in the presence of flowing and stagnant water and air currents (induced to complete cooling fans), suffer a serious deterioration of the steel by corrosion and biodeterioration, involving a variety of microorganisms. The timber is subjected to oxygen delignification under the effect of colonies of fungi and algae, as well as fiberglass reinforced polyester screens, which deteriorate due to colonies of aerobic and anaerobic bacteria e.g. sulfate reducers.

Furthermore, carbon steels corrode in the form of delamination due to sulfate reduction processes which induce the oxidation of iron, while the SS nails and screws undergoes localized corrosion, forming pits (Figure 17).

Figure 16: A humid corrosive environment in a geothermal field caused by steam and gases emission.

Figure 17: Corrosion in a cooling tower of a geothermal power plant.

The deterioration by microorganisms capable of living in these conditions is one of the processes that have provided more information to the study of corrosion induced by microorganisms. In "Cerro Prieto", for example, have been isolated and studied various bacteria capable of growing even at temperatures of 70 ° C under conditions of low nutrient concentrations, while in the geothermal field of «Azufres» bacteria have been isolated to survive at temperatures of 105 °C and pressures of downhole (Figure 18).

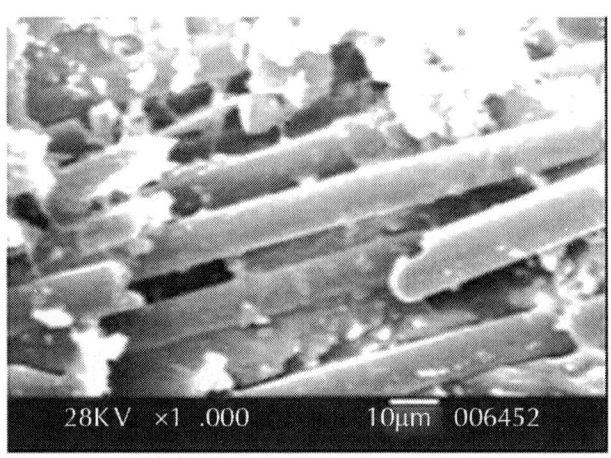

Figure 18: Biodeterioration of polyester polymeric matrix in a fiberglass screen exposed at geothermal temperatures.

CORROSION IN THE PAPER INDUSTRY

Corrosion of the infrastructure used in the pulping and paper industry, is another serious problem for corrosion specialists. The wide experience, gathered from cases of corrosion in the various infrastructure components of the paper industry, has provided an extensive literature on mechanisms, types and control of corrosion in this environment.

In the early 60's of last century, when the continuous digester process was adopted, the paper industry had limited knowledge about caustic embrittlement. Currently, it is known that the digesters are subjected to caustic levels and temperatures too close to the fracture caustic range where the total relieves of stresses in the material are essential. To elucidate the mechanism of this phenomenon, it was necessary to conduct serious investigations, which subsequently provide solutions to the problem of corrosion and caustic embrittlement. Technology in the paper industry has evolved over the last forty years and in parallel we can talk about the solution of corrosion problems in different parts of its infrastructure. Components with high failure rate due to corrosion are those built of bronze, SS, cast iron. Corrosion occurs in the papermaking machinery, where the white water equipment is subjected to an aggressive environment. The metal surfaces are exposed to immersion in this water; to steam that promotes the formation of cracks, which favor the deposit of pulp and other compounds. CS undergoes rapid uniform corrosion, while the copper alloys and SS (austenitic UNS S30400 L: 18% Cr8% Ni, UNS S31600 L: 16% Cr10% Ni 2% Mo) develop localized pitting corrosion. In the mill bleach plants the pulp equipment has traditionally been made of SS which has good general corrosion resistance and weldability. The use of chlorine gas (Cl_2) and oxygen in the bleach plant and pulp bleaching, favors a very aggressive oxidant and SS, as type 317 L (18% Cr14% Ni3.5% Mo). However, in the last 25 years the environment in these plants has become much more corrosive due to the wash systems employed for the paper pulp, which increased the emission of oxidizing and corrosive gases; so type "317 L" SS is not resistant and has a shorter service life. Many mills in the paper industry have opted for the use of high-alloy SS, nickel (Ni) and titanium (Ti), for better corrosion resistance in these particular environments. In general, SS exposed to corrosive environment of bleach plants are benefited by the share of chromium, nickel and molybdenum as alloying elements, which increase their resistance to the initiation of pitting and crevice corrosion. The addition of nitrogen (N) increases its resistance to pitting corrosion, particularly when it contains molybdenum (Mo). Furthermore, to avoid waste of elements such as carbon (C), where a concentration greater than 0.03%, can cause sensitization at affected by heat areas in the solder, causing the SS to be less resistant to corrosion. Other waste elements, such as phosphorus (P) and sulfur (S) can cause fractures in the hot steel, formed in the metal welding area. The corrosive environment of bleach plants contain residual oxidants such as chlorine (Cl_2) and chlorine dioxide (ClO_2), these are added to resists the effects of temperature and acidity, maintaining a very aggressive environment.

Corrosion also occurs in the pulping liquor facilities by sulfites, chemical recovery boilers, suction rolls and Kraft pulping liquors. The Kraft process is the method of producing pulp or cellulose paste, to extract the wood fibers, necessary for the manufacture of paper.

The process involves the use of sodium hydroxide (NaOH) and sodium sulfite (Na_2SO_3) to extract the lignin from wood fibers, using large high pressure digesters. High strength is obtained in the fiber and methods for recovery of chemicals explain the popularity of the Kraft process. The black liquor separated, is concentrated by evaporation and burned in a recovery boiler to generate high pressure steam, which can be used for the plant steam requirements for the production of electricity. The inorganic portion of the liquor is used to regenerate sodium hydroxide and sodium sulfite, necessary for pulping. Corrosion of metals in the facilities used in this process may occur during the acid pickling operation for the removal of carbonate incrustations on the walls and black liquor pipe heaters. It has been found that SS 304 L presents fracture failure and stress corrosion. In the recovery processes of chemical reagents, known as stage re alkalinization, metals can fail due to caustic embrittlement or corrosion-erosion under conditions of turbulent flow. Corrosion also occurs in the equipment used for mechanical pulping, such as stress corrosion cracking, crevice corrosion, cavitation and corrosion-friction.

REFERENCES

1. N. Acuña, B. Valdez, M. Schorr, G. Hernández-Duque, of. Effect, Biofilm. Marine, Fatigue. on, of. Resistance, Austenitic. an, Steel Stainless, Corrosion Reviews, 222101114United Kingdom 2004

2. M. Carrillo, B. Valdez, M. Schorr, L. Vargas, L. Álvarez, R. Zlatev, M. Stoytcheva, Actinomyces. In-vitro, biofilm. israelii's, on. I. U. D. development, surfaces. copper, Vol. Contraception, 32612642010

3. Carrillo Irene, Valdez Benjamin, Zlatev Roumen, Stoycheva Margarita, Schorr Michael, and Carrillo Monica, Corrosion Inhibition of the Galvanic Couple Copper-Carbon Steel in Reverse Osmosis Water, Research Article, Hindawi Publishing Corporation, International Journal of Corrosion, Volume2011Article ID 856415

4. 2010A. Garcia, B. Valdez, M. Schorr, R. Zlatev, A. Eliezer, J. Haddad, of. Assessment, marine, corrosion. fluvial, steel. of, Journal. aluminium, Marine. of, Engineering, Vol. Technology, 1839

5. Garcia Inzunza Ramses, Benjamin Valdez, Margarita Kharshan,Alla Furman, and Michael Schorr, (2012Interesting behavior of Pachycormus discolor Leaves Ethanol Extract as a Corrosion Inhibitor of Carbon Steel in 1M HCl. A preliminary study.Research Article, Hindawi Publishing Corporation, International Journal of Corrosion, 2012Article ID 980654, 8 2012

6. B. G. Lopez, B. Valdez, S. , R. Zlatev, K. , J. Flores, P. , M. Carrillo, B. , M. Schorr, W. , Corrosion of metals at indoor conditions in the electronics manufacturing industryAnti-Corrosion Methods and MaterialsUnited Kingdom, 54N0. 6, 354 EOF359 EOFNoviembre 2007

7. B. G. Lopez, B. Valdez, S. , M. Schorr, W. , N. Rosas, G. , H. Tiznado, V. , G. Soto, H. , Influence of climate factors on copper corrosion in electronic equipment and devicesAnti-Corrosion Methods and MaterialsUnited Kingdom, 57N0. 3, 148 EOF152 EOF2010

8. Lopez, Gustavo, Hugo Tiznado, Gerardo Soto Herrera, Wencel De la Cruz, BenjaminValdez, Miguel Schorr, Zlatev Roumen,2011Use of AES in corrosion of copper connectors of electronic devices and equipments in arid and marine environmentsAnti-Corrosion Methods and Materials58Iss: 6, 331336

9. Lopez, Badilla Gustavo, Benjamin Valdez Salas and Michael Schorr Wiener, Analysis of Corrosion in Steel Cans in the Seafood Industry on the Gulf of California, Materials Performance, Vol.45257April 2012

10. M. Navarrete, M. Ballesteros, J. Sánchez, B. Valdez, G. Hernández, in. a. Biocorrosion, power. geothermal, Materials. plant, April. Performance, 1999385256USA.

11. M. Schorr, B. Valdez, R. Zlatev, M. Stoytcheva, Corrosion. Erosion, Phosphoric. in, Production. Acid, Performance. Materials, Jan, 20105015659USA

12. M. Stoytcheva, B. Valdez, R. Zlatev, M. Schorr, M. Carrillo, Z. Velkova, Induced. Microbially, in. Corrosion, Mineral. The, Industry. Processing, Materials. Advanced, Research, 2010957376Trans. Tech publications, Switzerland

13. Raicho Raichev, Lucien Veleva y Benjamín Valdez, Corrosión de metales y degradación de materiales.Principios y prácticas de laboratorio. Editorial UABC, 978-6-07775-307-02009pp

14. S. N. Santillan, S. B. Valdez, W. M. Schorr, R. A. Martinez, S. J. Colton, of. Corrosion, heat. the, zone. affected, stainless. of, weldments. steel, Methods. Anti-Corrosion, United. Materials, Vol. Kingdom, 4180 EOF184 EOF2010

15. M. Schorr, B. Valdez, R. Zlatev, M. Stoytheva, N. Santillan, Ore. Phosphate, For. Processing, Acid. Phosphoric, Classical. Production, Novel. And, Mineral. Technology, Processing, Metallurgy. Extractive, Vol, 31251292010

16. Valdez, B., Guillermo Hernandez-Duque, Corrosion control in heavy-duty diesel engine cooling systems, CORROSION REVIEWS Vol.Nos. 2-4, 2452601995

17. Salas. B. Valdéz, Beltrán. Miguel, L. Rioseco, N. Rosas, J. A. Sampedro, G. Hernandez, M. Quintero, Corrosion control in cooling towers of geothermoelectric power plants. Corrosion Reviews, 142372521996England.

18. B. Valdez, N. Rosas, J. Sampedro, M. Quintero, J. Vivero, G. Hernández, of. Corrosion, concrete. reinforced, the. of, Colorado. Rio, aqueduct. Tijuana, Performance. Materials, May, 1999388082USA.

19. B. Valdez, N. Rosas, J. Sampedro, M. Quintero, J. Vivero, of. Influence, sulphur. elemental, corrosion. on, carbon. of, in. steel, environments. geothermal, Reviews. Corrosion, Vol, Nos. 3- 4, 167180October 1999England

20. S. B. Valdez, R. Zlatev, K. , M. Schorr, W. , N. Rosas, G. , Dobrev. M.Ts, I. Monev, Rapid. Krastev, for. method, protection. corrosion, of. V. C. I. determination, Anti. Films-Corrosion, Methods, United. Materials, Vol. Kingdom, 6362366Noviembre 2006

21. B. Valdez, M. Carrillo, R. Zlatev, M. Stoytcheva, M. Schorr, J. Cobo, T. Perez, J. M. Bastidas, Influence of Actinomyces israelii biofilm on the corrosion behaviour of copper IUD, Anti-Corrosion Methods and Materials, United Kingdom, 55N0. 2, 55-59, 2008

22. B. Valdez, M. Schorr, M. Quintero, M. Carrillo, R. Zlatev, M. Stoytcheva, J. Ocampo, Corrosion, at. scaling, Prieto. Cerro, Field. Geothermal, Methods. Anti-Corrosion, United. Materials, Vol. Kingdom, N0. 1, 28 EOF34 EOF2009

23. B. Valdez, M. Schorr, Control. Corrosion, The. in, Industry. Desalination, Materials. Advanced, Research, 2010952932Trans. Tech publications, Switzerland

24. B. Valdez, M. Schorr, M. Quintero, R. García, N. Rosas, effect. The, climate. of, on. change, durability. the, engineering. of, in. materials, hydraulic. the, infrastructure, An overview. Corrosion Engineering Science and Technology45134412010

25. B. Valdez, M. Schorr, A. So, A. Eliezer, Natural. Liquefied, Regasification. Gas, Materials. Plants, M. A. T. E. R. I. A. L. S. P. E. R. F. O. R. M. A. N. C. E. Corrosion, Vol, 126468December 2011

26. O. L. Vargas, S. B. Valdez, M. L. Veleva, K. R. Zlatev, W. M. Schorr, G. J. Terrazas, of. Corrosion, at. silver, conditions. indoor, assembly. of, in. processes, microelectronics. the, Anti. industry-Corrosion, Methods, United. Materials, Vol. Kingdom, N0. 4, 218 EOF225 EOF2009

27. L. Veleva, B. Valdez, G. López, L. Vargas, J. Flores, Corrosion. Atmospheric, Electro. of-Electronics, in. Metals, Indoor. Urban-Desert, Environment, Corrosion of Electro-Electronics Metals in Urban-Desert Indoor Environment. Corrosion Engineering Science and Technology4321491552008

Advances in Asset Management Techniques: An Overview of Corrosion Mechanisms and Mitigation Strategies for Oil and Gas Pipelines

Chinedu I. Ossai

Production Planning Department, Overall Forge Pty Ltd, 70 R W Henry Drive, Ettamogah near Albury, P.O. Box 5275, Albury, NSW 2640, Australia

ABSTRACT

Effective management of assets in the oil and gas industry is vital in ensuring equipment availability, increased output, reduced maintenance cost, and minimal nonproductive time (NPT). Due to the high cost of assets used in oil and gas production, there is a need to enhance performance through good assets management techniques.

This involves the minimization of NPT which accounts for about 20–30% of operation time needed from exploration to production. Corrosion contributes to about 25% of failures experienced in oil and gas production industry, while more than 50% of this failure is associated with sweet and sour corrosions in pipelines. This major risk in oil and gas production requires the understanding of the failure mechanism and procedures for assessment and control. For reduced pipeline failure and enhanced life cycle, corrosion experts should understand the mechanisms of corrosion, the risk assessment criteria, and mitigation strategies. This paper explores existing research in pipeline corrosion, in order to show the mechanisms, the risk assessment methodologies, and the framework for mitigation. The paper shows that corrosion in pipelines is combated at all stages of oil and gas production by incorporating field data information from previous fields into the new field's development process.

INTRODUCTION

The oil and gas industry is an asset intensive business with capital assets ranging from drilling rigs, offshore platforms and wells in the upstream segment, to pipeline, liquefied natural gas (LNG) terminals, and refineries in the midstream and downstream segments. These assets are complex and require enormous capital to acquire. An analysis of the five major oil and gas companies (BP, Shell, ConocoPhillips, Exxonmobil, and Total) shows that plant, property, and equipment on average accounts for 51% of the total assets with a value of over $100 billion [1]. Considering the huge investment in assets, oil and gas companies are always under immense pressure to properly manage them. To achieve this involves the use of different optimization strategies that is aimed at cost reduction and improved assets reliability [2].

Due to the growth in the demand of oil and gas around the world, companies are developing new techniques to reach new reservoirs in the offshore and onshore arena [3]. This is putting pressure on most of the facilities with the attendant cost of maintenance soaring [1]. The continuous utilization and the ageing of facilities have resulted in record failures in the oil and gas plants. Research shows that between 1980 and 2006, 50% of European, major hazards of loss containment events arising from technical plants failures were primarily due to ageing plants mechanism caused by corrosion, erosion, and fatigue [4, 5].

A study shows that corrosion cost in US rose above 1$ trillion in 2012 accounting for about 6.2% of GDP hence, the largest single expense in the economy [6]. In the oil and gas company, corrosion accounts for over 25% of assets failure [7] and is found to be prevalent in every stage of the production cycle. Oxygen which plays a dominant role in corrosion is normally present in producing formation water. During drilling operation, drilling mud can corrode the well casing, drilling equipment, pipeline, and the environment. Water and CO_2 produced or injected for secondary recovery can cause severe corrosion of completion strings, while the acids used to reduce formation damage around the well or to remove scale can attack metals [8]. The formation water and injected water used for the oil recovery are a potential source of pipeline corrosion during transportation of the oil from the wells to the loading terminals. Mechanical static equipment like valves, tanks, vessels, separators, and so forth are susceptible to a different kind of corrosion however, pipelines are more prone to corrosion due to the presence of CO_2, H_2S, H_2O, bacteria, sand, and so forth in the fluid.

Owing to the increasing cost of pipeline corrosion management in the oil and gas industries [1], operators are becoming more concerned about corrosion management planning at all phases of production. Corrosion information from existing field data is being incorporated into design information for new oil and gas field [9, 10] in a bid to develop appropriate corrosion management methodologies that will enhance the design life of the pipelines and optimize production. To reduce the risk of microbiologically influenced Corrosion (MIC) and other associated corrosions like stress corrosion cracking (SCC), hydrostatic testing of carbon steel pipes should be carried out in such a manner that enhances the future pipeline service conditions by using the right source of water, ensuring proper degree of filtration, ensuring limited exposure period to temperature and eliminating air packets [11]. Though bacteria in the biofilm are responsible for pitting of a pipeline in a MIC however, the impact of the flow velocity of the constituent fluid influences the mass transfer rate thereby affecting the biofilm formation, hence, inhibiting the activities of sulphate reducing bacteria, (SRB) present in the fluid [12]. This flow attribute has significant impact in MIC in oil and gas pipeline.

Considering the fact that the CO_2 and H_2S induced corrosion rate can reach up to 6 mm/yr and 300 mm/yr, respectively, [13] in oil and gas pipelines, sophistication in inspection and monitoring techniques is therefore necessary for quick mitigation. The increased trend in in-line inspection and online data acquisition has helped in quicker data acquisition, analysis, and decision making regarding corrosion in pipelines. The enhanced research knowledge of the behaviour of these corrodents (CO_2 and H_2S, acetic acid, etc.) at different operating conditions [14–17] has given rise to numerous mechanistic, statistical, and empirical models [18–23] which have contributed immensely in the inspection and monitoring, selection of inhibitors, and materials selection for pipelines design.

Since corrosion is a dominant factor contributing to failures and leaks in pipelines [24], to aid industry experts in managing the integrity of pipelines therefore involves a layout of the developments in the management strategies. This involves the recognition of the conditions contributing to the corrosion incident and identifying effective measures that can be taken to mitigate against them. To facilitate best practices in pipeline integrity management therefore, requires a framework that utilizes good policies and procedures in inspection, data collection, and interpretation for corrosion control.

OVERVIEW OF CORROSION

Corrosion is a naturally occurring phenomena commonly defined as the deterioration of a substance (usually metal) or its properties because of a reaction with its environment [25]. Corrosion of materials is inevitable due to the fundamental need of lowering of Gibbs energy [26]. Every material is trying to achieve a lower energy state hence the ability to corrode in order to get to a low energy oxide state. Though this is the case with all materials, the major focus of experts however, is to achieve an equilibrium position between the materials and the environment thereby controlling corrosion.

Modern corrosion science has its roots in electrochemistry and metallurgy. Whereas electrochemistry contributes to the understanding of materials via corrosion, metallurgy provides information about the behaviour of the material and their alloys hence provide a medium for combating the degradation on them. The type of corrosion mechanism and its rate of attack depend on the nature of the environment (air, soil, water, etc.) in which the corrosion takes place. Whereas some environmental condition can help to mitigate the rate of corrosion, others help to increase it hence, industrial wastes and products can either be corrosion inhibitor or catalyst. For instance, CO_2, H_2S, temperature, mass flow rate, pH, formation water, and so forth contribute in no small measure to the rate of corrosion in oil and gas pipeline [14, 16, 17, 27]. The existence of anodic cathodic sites on the surface of a piece of metal implies that the difference in electrical potential is found on the surface. This potential difference has the tendency of initiating corrosion. If an oil and gas pipeline passes through a zone of clay soil (where the oxygen concentration is low) to gravel (where the oxygen concentration is high), the part of the pipeline in contact with the clay becomes anodic and suffers damage. Though this problem is extensively addressed with the cathodic protection [26], concentration cell may also be formed where there are differences in metal ion concentration.

Although most metals are crystalline in form, they generally are not continuous single crystal but rather are collections of small grains of domains of localized order in which microcrystal forms as the liquid cools and solidifies. In the final states, the crystals have different orientation with respect to one another. The edge of the domain form grain boundaries which are an example of planar defects in metal. These defects are usually sites of chemical reactivity. The boundaries are also weaknesses, the places where stress corrosion cracking begins. The metallic surface exposed to an aqueous electrolyte usually possesses site for oxidation (anodic reaction) that produces electrons in the metal and reduction (cathodic reaction) that consumes the electrons produced by the anodic reaction [25, 26]. These sites make up a corrosion cell. The anodic reaction (Figure1) involves the dissociation of metal to form either soluble ionic product or an insoluble compound of metal usually an oxide. For cathodic reaction (Figure 2), oxygen gas generated could be reduced or water is reduced to produce hydrogen gas. The simultaneous reaction of the anodic and cathodic reactions produces the electrochemical cell.

Anodic process

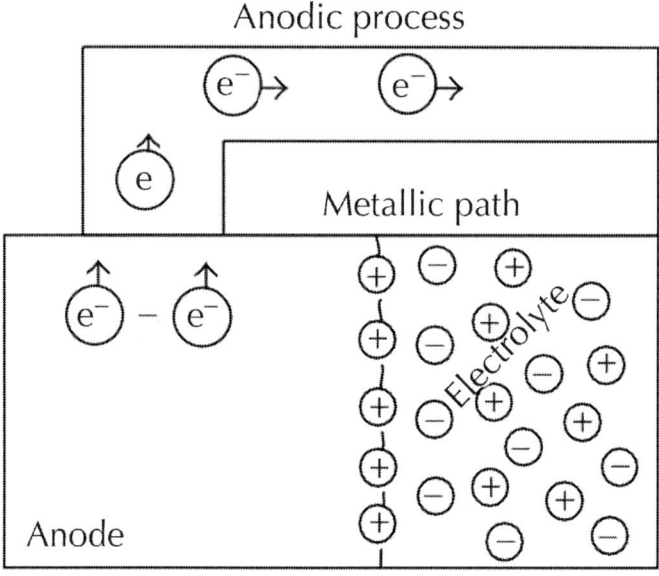

Figure 1: Anodic process.

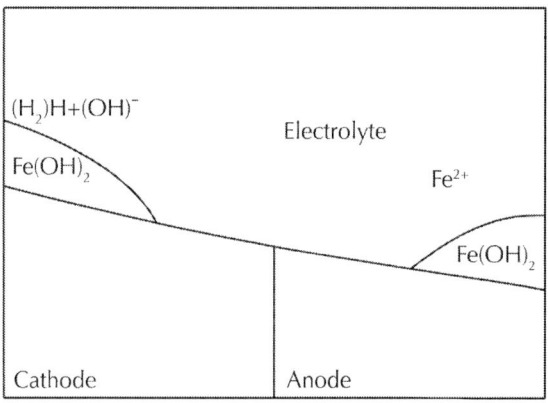

Figure 2: Cathodic process.

In completely oxygen-free water, the cathodic reaction that takes place is the reaction of hydrogen ion to form hydrogen gas as shown in (1):

$$2H^+ + 2e^- \longrightarrow H_2 \text{ (g)} \tag{1}$$

When significant amounts of oxygen are present in the system, the cathodic reaction that takes place is shown in (2):

$$2H^+ + \frac{1}{2}O_2 + 2e^- \rightarrow H_2O \tag{2}$$

The hydrogen ion is present in water due to the ubiquitous dissolution of water into hydroxyl ions as shown in (3):

$$2H_2O \longrightarrow 2H^+ + 2(OH) \tag{3}$$

In the anode, there is a dissociation of iron to form a ferrous ion as shown in (4).

$$Fe \longrightarrow Fe^{2+} + 2e \tag{4}$$

The ferrous ion will react with the hydroxyl ion to form insoluble ferrous hydroxide as shown in (5):

$$Fe^{2+} + 2(OH)^- \rightarrow Fe(OH)_2 \tag{5}$$

The anodic and cathodic reactions that take place in a neutral and alkaline condition is shown in Figure 3.

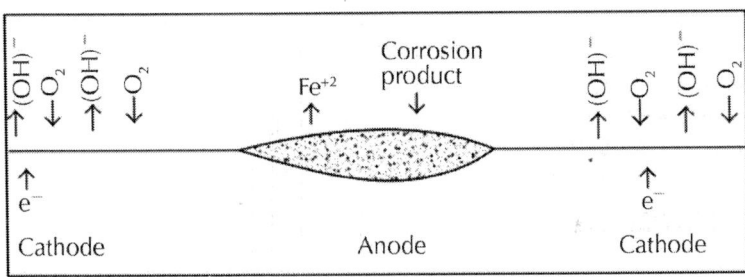

Figure 3: Neutral and Alkaline condition of a corrosion process.

The cathodic reaction is as follows:

$$\frac{1}{2}O_2 + H_2O + 2e^- \rightarrow 2(OH) \tag{6}$$

while the anodic reaction is the same as (4).

For an anodic condition, the cathodic and anodic reactions are represented in Figure 4.

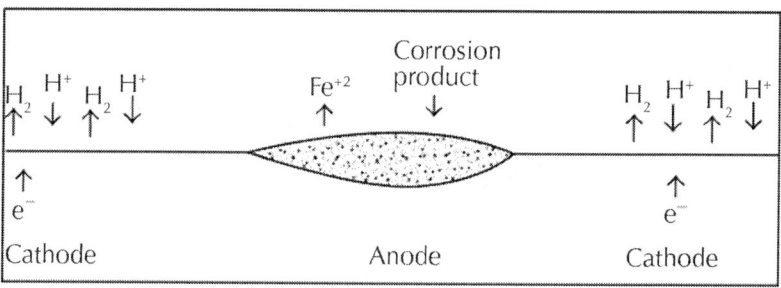

Figure 4: Acidic Condition of a Corrosion process.

The cathodic reaction equation is shown in (1), while the anodic equation is shown in (4).

In a deoxygenated solution, the hydrogen reaction combines with the others to yield the net corrosion reaction shown in (7):

$$Fe + 2H_2O \longrightarrow Fe(OH)_2 + H_2 \,(g) \tag{7}$$

In oxygenated aqueous systems, the oxygen reduction leads to a slightly different net corrosion reaction as shown in (8):

$$Fe + H_2O + \frac{1}{2}O_2 \rightarrow Fe(OH)_2 \tag{8}$$

Whereas in deoxygenated solution, hydrogen is evolved, in oxygenated system, oxygen is consumed. The evolved hydrogen acts as a catalyst for the formation of magnetite (Fe$_3$O$_4$) in deoxygenated water. Experiment shows that hydroxide readily decomposes into magnetite in deoxygenated water above 100°C [28] as indicated in (9):

$$3Fe(OH)_2 \longrightarrow Fe_3O_4 + H_2 (g) + 2H_2O \tag{9}$$

The net corrosion reaction with the magnetite as the final product is shown in (10):

$$3Fe + 4H_2O \longrightarrow Fe_3O_4 + 4H_2 (g) \tag{10}$$

In oxygenated solution, the ferrous oxide (Fe^{2+}) does not immediately precipitate out since it rapidly oxidizes to ferric oxide (Fe^{3+}), as a result, insoluble iron hydroxide is formed which is converted to hematite as shown in (11):

$$Fe + (OH)_2 + \frac{1}{2}H_2O + \frac{1}{4}O_2 \rightarrow Fe(OH)_3 \tag{11}$$

The ferric oxide (Fe^{3+}) is converted to magnetite according to (12)

$$2Fe(OH)_3 + Fe(OH)_2 \longrightarrow Fe_3O_4 + 4H_2O \tag{12}$$

Cathodic and anodic sites could be built as a result of variation in environmental conditions, metallic microstructure variation, and variation in environmental concentration of oxygen at different points of a metal [26]. At the anodic sites, the dissolution of metallic ions in the electrolyte brings about the flow of electrons between the corroding anodes and non-corroding cathodes. The spontaneous nature of the corrosion however, depends on the rate of flow of these electrons.

Though establishing the tendency for corrosion is necessary, however, it is more important to determine the rate of corrosion. This is because a particular metal or alloy may be prone to corrosion in an environment but at a very low rate, in which it will not be a problem [26]. To understand the rate of corrosion however, requires the knowledge of the role of primary environment and metallurgical variables, underlying mechanism of corrosion, and synthesis of information to account for effects of the parameters.

MECHANISMS OF CORROSION IN OIL AND GAS PIPELINES

Fluid flowing from oil and gas pipelines has a combination of chemicals including CO_2, H_2S, organic acids, bacteria, sand, and water. These constituents are among the major causes of corrosion in pipeline. The CO_2 dissolves in the presence of water to form an acidic oxide which reacts with iron. This type of corrosion is referred to as sweet corrosion. This is responsible for most types of general corrosion in oil and gas pipeline. Sour corrosion occurs when H_2S in the excess of 100ppm is present in the oil and gas, causes corrosion in the pipeline, and predominantly causes pitting [26, 29].

CO_2 present in oil and gas will dissolve in water to produce carbonic acid (H_2CO_3) [23, 27]. This acid dissolves steel to produce iron carbonate and hydrogen as shown in (13). This reaction takes place at the cathode:

$$Fe + H_2CO_3 \longrightarrow FeCO_3 + H_2\ (g) \tag{13}$$

Despite the weakness of carbonic acid it is extremely corrosive to carbon steel. The chemical reactions above form the iron carbonate films. Depending on the condition during the formation, these films can be protective or non-protective at the anode, iron dissolves as shown in (4). The presence of CO_2 acts as a catalyst increasing the hydrogen evolution thereby increasing the corrosion rate of carbon steel in aqueous solution [27]. The carbonic acid (H_2CO_3) either serves as an extra source of H^+ or is reduced directly according to (14) and (15):

$$2H^+ + 2e^- \longrightarrow H_2\ (g) \tag{14}$$

$$2H_2CO_3 + 2e^- \longrightarrow H_2\ (g) + 2HCO_3 \tag{15}$$

The dissolved iron concentration will increase until Fe^{2+} is the same as the precipitation rate of $FeCO_3$ [30]. When Fe^{2+} is released in the corrosion process, the double amount of bicarbonate forms according to (16):

$$Fe + 2H_2CO_3 \longrightarrow Fe^{2+} + H_2 + 2HCO_3 \qquad (16)$$

The pH increases until bicarbonate and carbonate becomes so high that solid $FeCO_3$ precipitates [30] as shown in equation (17):

$$Fe^{2+} + 2HCO_3^- \longrightarrow FeCO_3 (s) + H_2CO_3 \qquad (17)$$

When all the ferrous ions produced by corrosion precipitates as iron carbonate ($FeCO_3$), the pH remains constant and the overall reaction becomes as the state in (13).

In order to control the rate of corrosion on the pipeline, there should be passivity. Passivity is the condition existing on a metal surface because of the presence of protective film. When protective film is formed on the metal surface, it forms a coat which hinders further corrosion action on the material [26, 31]. The structure of the passive film (magnetite) formed on low carbon steel oxidizes in high temperature and has two distinct layers on the steel. The inner layer is compact and adheres well to the steel and has uniform thickness. The outer layer is a porous mass of individual crystal that would flake off the steel in some place and very nonuniform in thickness (Figure 5). This protective film is removed from the surface of the pipeline through erosion, dissolution, and turbulence resulting in more corrosion. The possible mechanisms resulting in the removal of the protective film are a follows:

(i)   Dissolution or removal of protective layer by hydrodynamic shear stress occurs when the shear stress is greater than the bonding force between the film and the substrate. This is a function of a mechanical process of erosion caused by the multiphase flow regime in pipeline [19, 32].

(ii) In distributed flow condition, local near wall density of turbulence helps to remove the protective film. This disruption to the mass transfer boundary layer results in an enhanced corrosion rate [32,33].

(iii) Dissolution of film which is controlled by mass transfer. Thus the breakaway velocity may reflect conditions where the dissolution rate of the film is greater than the growth rate of the film [34].

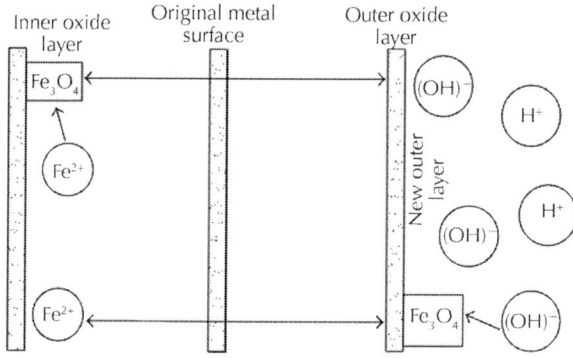

Figure 5: Schematic of magnetite double layer showing oxide formation locations.

The breakdown of protective film leads to the formation of localized corrosion that results in some of the major sources of corrosion failures like pitting, crevice, intergranular, and stress corrosion [12, 26, 35, 36]. The predominant breakdown processes are electrochemical and mechanical. Mechanical breakdown occurs when the protective film is ruptured as a result of stress or abrasive wear while, the electrochemical breakdown is a function of chemical reaction between the fluid constituent and the steel.

Types of Corrosion in Oil and Gas Pipeline

The primary chemical components that cause corrosion reaction to occur in pipeline are oxygen, acidic sulphur, and acidic chloride that dissolves in the water in the pipeline. The mechanism present in a given piping system varies according to the fluid composition, service location, geometry, temperature and so forth. In all cases of corrosion, the electrolyte must be present for the reaction to occur.

Internal Corrosion

Internal corrosion has become an increasing problem in most oil and gas pipelines as water cuts have increased and previously oil wet pipe surfaces have become water wet (providing the electrolyte for the corrosion cell) and as bacterial activities increases in the production system. Internal corrosion is the largest cause of pipeline failure in oil and gas industries [24] through different forms of corrosion like microbiologically influenced corrosion (MIC), erosion (flow enhanced) corrosion, under deposit (concentration cell) corrosion and so forth.

Erosion-Corrosion

The erosion-corrosion mechanism increases corrosion reaction rate by continuously removing the passive layer of corrosion products from the wall of the pipe. The passive layer is a thin film of corrosion product that actually serves to stabilize the corrosion reaction and slow it down. As a result of the turbulence and high shear stress in the line, this passive layer can be removed causing the corrosion rate to increase [37]. The erosion-corrosion is always experienced where there is high turbulence flow regime with significantly higher rate of corrosion than just corrosion or erosion in pipeline [38]. In a multiphase flow regime with a fully developed turbulent flow, bubbles development and collapse have been attributed to changes in mass transfer coefficient and an eventual increase in CO_2 corrosion in pipeline [34].

Under Deposit Corrosion

The under deposit mechanism can increase the corrosion reaction rate by causing a localized chemical concentration which results in pitting of the metal surface under solid deposits. These deposits appear to be composed of a corrosion product matrix with entrapment of formation solids, sand, and iron sulphide. The rate of corrosion under this mechanism is significantly lower than erosion-corrosion mechanism.

Microbiologically Induced Corrosion (MIC)

This type of corrosion is caused by bacterial activities. The bacteria produce waste products like CO_2, H_2S and organic acids that corrode the pipes by increasing the toxicity of the flowing fluid in the pipeline. Some bacteria like sulphate removing bacteria (SRB) consume hydrogen that is a product in a standard corrosion reaction process. This activity causes the existing corrosion rate to increase in an attempt to reach reaction equilibrium by replacing the hydrogen consumed by bacteria. Bacteria also accumulate on the pipe walls, creating deposits and under deposit corrosion. MIC is recognized by the appearance of black slimy waste material or nodules on the pipe surface as well as pitting of the pipe wall underneath these deposits.

Pitting Corrosion

Pitting is classified as a localized attack that results in rapid penetration and removal of metal at small discrete area. The initiation of a pit occurs when electrochemical or chemical breakdown exposes a small local site on a metal surface to damaging species such as chloride ion. The site where pitting occurs is where there is an environmental variation in comparison to the entire metal surface. The combination of chlorine with H_2S results in localized pitting on steel [35]. This area of pitting which is usually the anode normally get highly degraded due to enormous electron transfer between the entire large area of the metal surface which is the cathode and small anode (the pitting site).

Crevice Corrosion

Crevice corrosion results when a portion of a metal surface is shielded in such a way that the shielded portion has limited access to the surrounding environment. Such surrounding environment contain, damaging corrosion species usually chloride ion. A typical example of crevice corrosion is the crevice found at the area between two metal surfaces in close contact with a gasket or another metal surface. The environment that eventually forms in the crevice is similar to that formed under the precipitated corrosion that covers a pit. An electrochemical corrosion cell is formed from the couple between the unshielded surface and the crevice interior exposed to an environment with a lower oxygen concentration compared with the surrounding medium. The concentration of being the anode of a corrosion cell and existing in an acidic, high-chloride environment where repassivation is difficult makes the crevice interior subject of corrosion attack.

Stress Corrosion Cracking (SCC)

Stress corrosion cracking (SCC) is a form of localized corrosion which produces cracks in metals by simultaneous action of a corrodent and tensile stress. It propagates over a range of velocities from 10^{-3} -10 mm/h depending upon the combination of alloy and environment involved. The geometry is such that if they grow to appropriate lengths, they may reach a critical size that results in a transition from the relatively slow crack growth rate associated with stress corrosion to fast crack propagation rates associated with purely mechanical failure. This transition happens when the stress intensity, which is a function of the geometry of the component including the crack size, reaches the fracture value for the material concerned. SCC in pipeline is a type of environmentally associated cracking (EAC). This is because the crack is caused by various factors combined with the environment surrounding the pipe. The most obvious identifying characteristic of SCC in pipeline is high pH of the surrounding environment, appearance of patches, or colonies of parallel cracks on the external of the pipe [39].

Top of the Line Corrosion (TLC)

This type of corrosion occurs due to the inability of corrosion inhibitors getting to the top of the pipeline (12 o'clock) thereby exposing it to corrodents. The inhibition effect is found to be predominant at the bottom of the line (6 o'clock), 9 o'clock and 3 o'clock where the flow of the oil or gas is taking place. This exposes the top of the line to concerted attack by the agents of corrosion with a resultant failure at some point. The primary factor that affects TLC is temperature which acts on the iron carbonate film formed. The combined effect of temperature fluctuation and condensation rate exposes the iron carbonate film to deterioration and consequently more corrosion. A study of the influence of gas flow rate on TLC shows that higher flow rate (which results in higher condensation rate) brings about more corrosion [40], while at a certain critical condensation rate, temperature and pH, TLC does not occur in gas pipelines [41]. The presence of acetic acid (HAc) has been found to enhance CO_2 TLC on carbon steel pipe, though at certain concentration level, HAc does not affect CO_2 TLC in carbon steel [42].

External Corrosion

External corrosion is caused by water penetrating the insulation system and is trapped between the insulation and the external pipe wall. The corrosion cell is fuelled by the continual supply of water and oxygen from the external sources. The main area where external corrosion is found is at the field applied weld insulation packs, but it can also be at any location where the galvanized insulation jacket has been punctured or torn. Weld pack insulations that are not well sealed allow water ingress making the weld packs to be wet. A fairly high temperature is needed to drive the corrosion mechanism, and the longer the mechanism has been active, the worst the damage will be. Therefore, the hottest and coldest lines in the field should have the highest likelihood for having an external corrosion problem.

CORROSION MANAGEMENT TECHNIQUES

Corrosion management is that part of the overall management system, which is concerned with the development, implementation, review, and maintenance of corrosion policy [7]. The corrosion policy, however, is a framework on which decision concerning corrosion issue in an industrial setting is based. This framework provides basic measures for risk determination via development of absolute risk control measures through planning, implementation, and control strategies. Corrosion management contributes to numerous benefits like statutory or corporate compliance with safety, health and environment policies, reduction in leaks, increased plant availability, reduced unplanned maintenance, and reduction in deferment costs [43]. To manage corrosion involves the utilization of a framework that will model the organization's policy through organizing, planning and implementing, measuring and reviewing, and auditing performance at all levels of execution as shown in Figure 6.

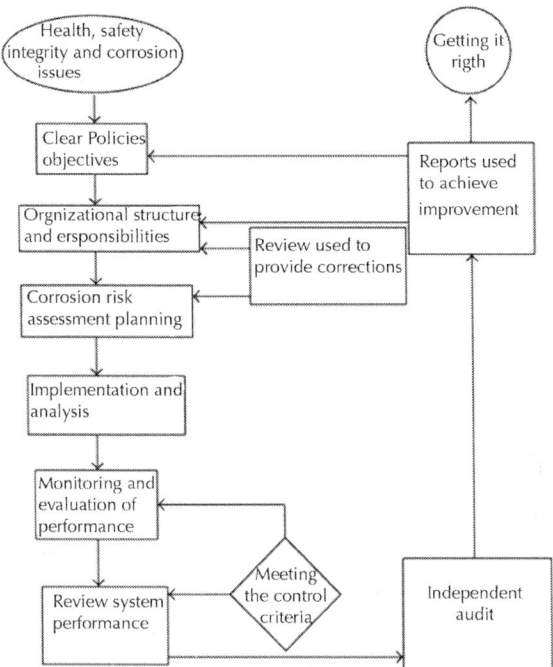

Figure 6: Corrossion management framework.

Corrosion Risk Assessment (CRA)

In planning for corrosion management, there is need for a formal identification of facilities that have the risk of degradation due to corrosion. The purpose of corrosion risk assessment is to rank facilities in order of their proneness to corrosion, identify options to remove, mitigate, or manage the risks. In order to manage corrosion risks, monitoring and inspection program will be incorporated in the overall activity schedule of an organization. The probability of failure is estimated based on the type of corrosion damage expected to occur on the component while the consequences of failure are measured against the impact of such a failure evaluated against a number of criteria. The criteria could include potential hazards to environment, risks associated with safety and integrity, or risk due to corrosion or inadequate corrosion mitigation procedure.

Typical of the risk-based procedure is the Failure Mode, Effect and Critical Analysis (FMECA) that ranks perceived risks in order of seriousness as shown in (18):

Criticaly (Risk)

$$=\text{Effect (Consequence)} * \text{Mode (Probable Frequency)}, \qquad (18)$$

where: failure Criticality is potential failures as examined in order to predict the severity of each failure effect in terms of safety, decreased performance, total loss of function and environmental hazards. Failure effect is potential failures assessed to determine the probable effect on process performance and the effects of the components on each other. Failure mode is the anticipated operational conditions used to identify most probable failure mode, the damage mechanism and likely locations.

Corrosion risk is the product of the probabilities of a corrosion-related failure and the consequences of such a failure [44]. The risk analysis of a pipeline is a measure of the probability of failure. The acceptable annual failure probability is dependent on the safety class [45] as shown in Table 1.

Table 1: Safety class and target annual failure probability

Safety class	Annual failure probability
High	$<10^{-3}$
Medium	$<10^{-4}$
Low	$<10^{-5}$

Corrosion risk assessment can be carried out on a group of components which are constructed from the same material and subject to the same operating condition or an individual component. In oil and gas pipelines, the risk is analyzed as either external or internal corrosion threat or environmental and operational threat. The remaining life of the pipeline is estimated against some established operational standards, while the rate of corrosion is correlated with the operating parameters of the oil and gas like CO_2, H_2S, temperature, pressure, flow rate, water cut and so forth. For effective corrosion assessment, the information concerning the operating condition of a facility will be maintained throughout the life cycle. The information is useful in formulating a corrosion risk assessment model that will be validated and modified with new assumptions overtime. For a non-stable process condition, detailed re-assessment will be required at least annually but a stable process with good historical data trend will need revalidation less frequently [7].

Risk Based Inspection (RBI)

In managing oil and gas pipelines against corrosion, RBI technique is used to develop an optimum plan for the execution of the inspection activities. RBI uses findings from corrosion risk assessment (CRA) or other risk analysis to plan physical inspection procedures. A risk-based approach to inspection planning will ensure that risk is reduced to as low as reasonably practicable. It will also optimize inspection schedule, focus effort on the most critical area, and identify the most appropriate methods of inspection [46]. Planning a risk-based analysis involves listing activities, task and other elements of a project, identifying the technical risks, develop a risk ranking factor scale for each activity, document results and identify potential risks reduction actions for evaluation by personnel [47].

Corrosion Monitoring

Corrosion inspection and monitoring are key activities in ensuring, pipelines integrity are maintained and corrosion mitigated [48]. The choice of corrosion control measure is a function of fluid composition, pressure, temperature, aqueous fluid corrosivity, facility, and technical culture inherent in an establishment. In monitoring and inspection of pipelines, data are collected to enhance corrosion control by way of predicting the remaining life and the suggestion of possible mitigation measures that will help to enhance serviceability will largely depend on the experience of the personnel. A thorough practice for corrosion management involves the monitoring of corrosion risks through proactive and reactive monitoring techniques. In management of pipeline corrosion in oil and gas industries, proactive technique which involves determination of the corrosion standpoint prior to failure is utilized. This involves in-line and on-line monitoring system. In this system, data which could enhance the knowledge of the rate of corrosion degradation are collected and steps are taken to prevent failure. In-line system cover the installation of devices directly into the pipeline like corrosion coupons, biostuds and so forth. These need to be extracted for analysis periodically. On-line monitoring techniques include deployment of corrosion monitoring devices either directly into the process or fixed permanently to the facility. These include electrical resistance (ER) probes, linear polarization resistance (LPR) probes, fixed ultrasonic (UT) probes, acoustic emission and so forth.

Whereas some corrosion monitoring techniques can be used for continuous monitoring, others are used for periodic monitoring. Corrosion monitoring techniques can either be direct or indirect parameter measure. This is summarized in Table 2.

Table 2: Summary of corrosion monitoring techniques

Direct method	Indirect method
Non-destructive inspection (NDI)	Biological counts
Material test coupons	Hydrogen probes
Electrical resistance (ER) probes	pH probes
Linear polarization resistance (LPR)	Specific ions
Elector-chemical impedance spectroscopy (EIS)	Temperature
Electro-chemical noise (EN)	conductivity
Galvanic current (GC)	Electrical potential monitor

Corrosion Mitigation Strategies

After corrosion risk assessment and data collection and analysis are completed, there is need for corrective action on the facility; this depends on the level of the deterioration experienced by facility. The approaches available for mitigating corrosion in pipeline includes, coating surfaces to act as a barrier or perhaps provide sacrificial protection, the addition of chemical specie to the environment to limit corrosion, alteration of alloy chemistry to make it more resistance to corrosion and utilization of alternative material [24].

Effective corrosion mitigation involves a good approach to assessment linked to inspection monitoring during initial design and re-evaluation of pipeline with respect to the selection of inhibitors. The summary of inhibitor selection for carbon steel pipeline at different risk categories is shown in Table 4.

Corrosion can be prevented or controlled by understanding the principle underlying corrosion process. This understanding has been the basis for the development of a number of corrosion prevention measures. The basic corrosion control measures are based on electrochemical driving force as shown in Pourbaix diagram in Figure 7. Table 3 shows the different pipeline corrosion mitigation strategies.

Table 3: Shows the different pipeline corrosion mitigation strategies

Mitigation strategy	Option	Remarks
Appropriate materials	Use of corrosion resistant alloys, non-metallic materials like Reinforced composite, thermoplastic-lined and polyethylene pipelines. Consider use of internally coated carbon steel pipeline systems (i.e., nylon or epoxy coated) with an engineered joining system.	(i) Non-metallic materials may be used as a liner or a free standing pipeline depending on the service conditions. (ii) Selection of appropriate material at construction and major refurbishment stage is necessary.
Chemical treatment	Corrosion inhibitors, biocides, oxygen scavengers, gas blanketing, vacuum deaeration	(i) The presence of small amounts of oxygen (parts per billion) or bacteria will accelerate corrosion. (ii) Provides a barrier between corrosive elements and the pipe surface
Coating and lining	Organic Coatings, metallic coatings, lining, cladding	Useful for internal and external corrosion prevention
Cathodic protection	Sacrificial anodes, impressed current systems, hybrid system	Need ability to monitor performance on-line.

Process control	Identify key parameters: pH, temperature, pressure, Flow rate, water chemistry, pH, chlorides, dissolved metals, bacteria, suspended solids, chlorine, oxygen, and chemical residuals	(i) Changes in operating conditions will influence the corrosion potential. Production information can be used to assess corrosion susceptibility based on fluid velocity and corrosivity (ii) Trends in dissolved metal concentration (i.e., Fe, Mn) can indicate changes in corrosion activity
Design detailing	Ensure ease of access and replacement: (i) Install valves that allow for effective isolation of pipeline segments from the rest of the system (ii) Install binds for effective isolation of in-active pipeline segments	Allows the effective suspension and discontinuation of pipeline segments: (i) Removes potential "deadlegs" from the gathering system (ii) Develop shut-in guidelines for the timing of required steps to isolate and lay up pipelines in each system

Table 4: Corrosion inhibitor risk categories

Risk category	Max inhibitor availability	Max expected uninhibited corrosion rate (mm/yr)	comments	Proposed category name
1	0%	0.4	Benign Fluid, corrosion inhibitor use not anticipated. Predicted metal loss accommodated by corrosion allowance	Benign
2	50%	0.7	Corrosion inhibitor probably required but with expected corrosion rates there will time be time to review the need for inhibition based on inspection data.	Low
3	90%	3	Corrosion inhibition required for majority of field life but inhibitor facilities need not be available from day one.	Medium

| 4 | 95% | 6 | High reliance on inhibition for operational life time. Inhibitor facilities most be available from day one to ensure success | High |
| 5 | >95% | >6 | Carbon steel and inhibition is unlikely to provide integrity for full field life. Select corrosion resistant material or plan for repair and replacement | Unacc-eptable |

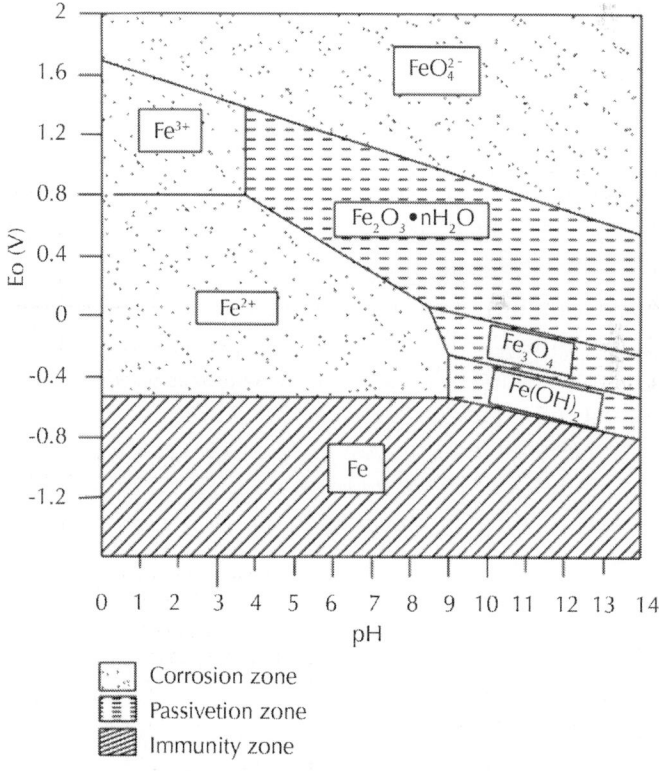

Figure 7: Pourbaix diagram of iron corrosion, passivity and immunity.

CONCLUSIONS

The prevalence of corrosion in oil and gas industry has resulted in enormous investment in technology to help combat impacts like loss of containment, leakages, death of personnel an environmental pollution. To this end, new oil and gas fields are developed using experiences generated from previous fields with similar characteristics. Efforts of design personnel at ensuring that the carbon steel materials are operating in the environment of immunity or passivity as shown in Pourbaix diagram (Figure 7) are yielding results via introduction of high corrosion resistant carbon alloys for pipelines. In other instances, specialized corrosion resistant materials have been used for lining the pipelines while reinforced composite and PVC materials have been utilized as alternative material for pipeline construction.

Advancements in inspection and monitoring techniques are also aiding corrosion experts in decision concerning the "when" and "how" pipelines are managed in a bid to optimize performance and cost. The proliferation of different empirical, statistical, and mechanistic prediction models for corrosion prediction is aiding personnel in managing the integrity of the pipelines through different mitigation strategies.

Finally, if pipeline corrosion which is a major contributor to nonproductive time (NPT) in oil and gas production will be reduced to the barest minimum, a corrosion management policy with a well-defined structure that includes responsibilities, reporting routes, practices, procedures, and resources has to be strictly followed in the oil and gas industries. The effectiveness of the policy will therefore depend on the willing of the leadership and commitment of other personnel at all ranks.

REFERENCES

1.   R. Nicholson, J. Feblowitz, C. Madden, and R. Bigliani, "The Role of Predictive Analytics in Asset Optimization for the Oil and Gas Industry-White Paper," 2010, http://www.tessella.com/wp-content/uploads/2008/02/IDCWP31SA4Web.pdf.

2.   J. Neelamkavil, "A review of Existing Tools and Their Applicability to Maintenance Management. Report # RR-285,"http://pdf. aminer.org/000/274/575/a_decision_support_system_for_ transmission_facility_maintenance.pdf.

3.   Oil and Gas Enhanced Production Services Industry to 2016—Enhanced Oil Recovery (EOR) Driving E&P Activity in Depleting Hydrocarbon Reservoirs, http://www.reportlinker. com/p0845623/Oil-and-Gas-Enhanced-Production-Services-Industry-to-Enhanced-Oil-Recovery-EOR-Driving-E-P-Activity-in-Depleting-Hydrocarbon-Reservoirs.html.

4.   Control of Major Accident Hazards, "Ageing Plant Operational Delivery Guide,"http://www.hse.gov.uk/comah/guidance/ ageing-plant-core.pdf.

5.   P. Horrocks, D. Mansfield, K. Parker, J. Thomson, T. Atkinson, and J. Worsley, "Managing Ageing Plant," http://www.hse.gov.uk/ research/rrpdf/rr823-summary-guide.pdf.

6.   "Cost of Corrosion to Exceed $1 Trillion in the United States in 2012—G2MT Labs -The Future of Materials Condition Assessment," http://www.g2mtlabs.com/2011/06/nace-cost-of-corrosion-study-update/.

7.   Review of Corrosion Management for Offshore Oil and Gas Processing, HSE OffshoreTechnology Report 2001/044, 2001.

8.   Corrosion in the Oil Industry(Oilfield review) Schlumberger,http:// www.slb.com/resources/publications/industry_articles/oilfield_ review/1994/or19940401_corrosion.aspx.

9.   B. Khajota, D. Sormaz, and S. Nesic, "Case-based reasoning model of CO_2 corrosion based on field data," CORROSION, 2007, paper no. 07553.

10. K. U. Raju, "Successful scale mitigation strategies in Saudi Arabian oil fields," in International Symposium on Oilfield Chemistry, The Woodlands, Tex, USA, April 2009, paper no. 121679. View at Scopus

11. A. Darwin, K. Annadorai, and K. Heidersbach, "Prevention of corrosion in carbon steel pipeline containing hydrostatic water-an overview," in CORROSION, March 2010, paper no. 10401.

12. J. Wen, T. Gu, and S. Nesic, "Investigation of the effects of fluid flow on SRB biofilm," inCORROSION, 2007, Paper no. 07516.

13. B. Hedges, H. J. Chen, T. H. Bieri, and K. Sprague, "A review of monitoring and inspection technique for CO_2 and H_2S corrosion in oil and gas production facilities: location, location, location," inCORROSION, 2006, paper no. 06120.

14. M. Singer, B. Brown, A. Camacho, and S. Neši , "Combined effect of carbon dioxide, hydrogen sulfide, and acetic acid on bottom-of-the-line corrosion," Corrosion, vol. 67, no. 1, 2011.

15. K. L. J. Lee and S. Nesic, "EIS investigation of CO_2/H_2S corrosion," in CORROSION, April 2004, paper no. 04728.

16. K. D. Ralston and N. Birbilis, "Effect of grain size on corrosion: a review," Corrosion, vol. 66, no. 7, pp. 0750051–07500513, 2010.

17. Y. Song, A. Palencsár, G. Svenningsen, J. Kvarekvål, and T. Hemmingsen, "Effect of O_2 and temperature on sour corrosion," Corrosion, vol. 68, no. 7, pp. 662–671, 2012.

18. A. Kale, B. H. Thacker, N. Sridhar, and C. J. Waldhart, "A probabilistic model for internal corrosion of gas pipelines," in Proceedings of the 5th Biennial International Pipeline Conference (IPC '04), pp. 2437–2445, Calgary, Canada, October 2004.

19. S. Nesic, J. Cai, and K.-L. J. Lee, "A multiphase flow and internal corrosion prediction model for mild steel pipeline," in CORROSION, 2005, Paper no. 05556.

20. W. Sun and S. Nesic, "A mechanistic model of H_2S corrosion of mild steel," in CORROSION, 2007, paper no. 07655.

21. X. Hu, V. D. Souza, A. Neville, and J. Well, "Prediction of erosion-corrosion in oil and gas- a systematic approach," in CORROSION, 2008, paper no. 08540.

22. X. Tang, C. Li, F. Ayello, J. Cai, and S. Nesic, "Effects of oil type on phase wetting transition and corrosion in oil-water flow," in CORROSION, NACE International, 2007, Paper no. 017170.

23. Y. Xian and S. Nesic, "A stochastic prediction model of localized CO_2 corrosion," in CORROSION, 2005, paper no. 05057.

24. CAPP, "Best Management Practices: Mitigation of Internal Corrosion in Oil Effluent Pipeline Systems," 2009, http://www.capp.ca/getdoc.aspx?DocId=155641&DT=PDF.

25. B. A. Shaw and R. G. Kelly, "What is corrosion?" Electrochemical Society Interface, vol. 15, no. 1, pp. 24–26, 2006.

26. J. Kruger, "Electrochemistry of Corrosion," 2001, http://electrochem.cwru.edu/encycl/art-c02-corrosion.htm.

27. V. Fajardo, C. Canto, B. Brown, and S. Nesic, "Effect of organic acids in CO_2 corrosion," inProceedings of the NACE International Conference and Exposition CORROSION, 2007, paper no. 07319.

28. P. S. Joshi, G. Venkateswaran, K. S. Venkateswarlu, and K. A. Rao, "Stimulated decomposition of $Fe(OH)_2$ in the presence of AVT chemicals and metallic surfaces—relevance to low-temperature feedwater line corrosion," CORROSION, vol. 49, no. 4, pp. 300–309, 1993.

29. R. Nyborg, "Controlling internal corrosion in oil and gas pipeline," Business Briefing-Exploration & Production: The Oil & Gas Review, no. 2, pp. 70–74, 2005.

30. A. Dugstad, E. Gulbrandsen, M. Seiersten, J. Kvarekval, and R. Nyborg, "Corrosion testing in multiphase flow, challenges and limitations," in CORROSION, 2006, paper no 06598.

31. R. N. Kig, "A review of fatigue crack growth in air and seawater," Offshore Technology Report OTH96 511, HSE, 1996.

32. A. Keating and S. Nesic, "Prediction of two-phase erosion-corrosion in bends," in Proceedings of the 2nd International Conference on CFD in Minerals and Processes Industries CSIRO, Melbourne, Australia, December 1999.

33. S. Nesic and J. Postlethwaite, "Relationship between the structure of disturbed flow and erosion-corrosion," Corrosion, vol. 46, no. 11, pp. 874–880, 1990.

34. H. Wang, W. Paul Jepson, J.-Y. Cai, and M. Gopal, "Effect of bubbles on mass transfer in multiphase flow," in CORROSION, 2000, paper no. 00050.

35. H. Fang, B. Brown, and S. Nešiæ, "Effects of sodium chloride concentration on mild steel corrosion in slightly sour environments," in CORROSION, vol. 67, no. 1, January 2011.

36. E. Mysara Mohyaldinn, N. Elkhatib, and C. Mokhtar Ismail, "A computational tool for erosion/corrosion prediction in Oil/Gas production facilities," in Proceedings of 3rd International Conference on Solid State Science & Technology (ICSSST '10), Kuching, Malaysia, December 2010.

37. Sh. Hassani, K. P. Roberts, S. A. Shirazi, J. R. Shadley, E. F. Rybicki, and C. Joia, "Flow loop study of NaCl concentration effect on erosion, corrosion, and erosion-corrosion of carbon steel in CO_2-saturated systems," in CORROSION, vol. 68, no. 2, February 2012.

38. A. A. Sami and A. A. Mohammed, "Study synergy effect on erosion-corrosion in oil and gas pipelines," Engineering and Technology, vol. 26, no. 9, 2008.

39. M. Baker Jr., "Stress Corrosion Cracking Study," 2004,http:// www.polyguardproducts.com/products/pipeline/TechReference/ SCC_Report-Final_Report_with_Database.pdf.

40. S. Olsen and A. Dugstad, "Corrosion under dewing conditions," in CORROSION, 1991, paper no. 472.

41. F. Vista and K. Alam, "Semi-empirical model for prediction of top-of-the-line corrosion risk," inCORROSION, 2002, paper no. 02245.

42. C. Mendex, M. Singer, A. Camacho, S. Hernndez, and S. Nesic, "Effect of acetic acid pH and MEG on CO_2 top of the line corrosion," in CORROSION, 2005, paper no. 05278.

43. D. Storey, "A Service Company's Experience with Pipeline Integrity Management," 2004,http://www.roseninspection.net/ MA/papers/2004-11_PipelineIntegrityManagement.pdf.

44. P. O. Gartland and J. Roy, "Application of internal corrosion modelling in risk assessment of pipeline," in CORROSION, 2003, paper no. 03179.

45. Det Norske Veritas (DNV RPG 101), Recommended Practice DNV-RP-101: Corroded Pipelines, 2010.

46. Det Norske Veritas (DNV RPG 101), "Risk Based Inspection of Topsides Static Mechanical Equipment, 2001".

47. P. K. John and L. D. John, "Risk factor analysis—a new qualitative risk management tool," inProceedings of the Project Management Institute Annual Seminar & Symposium, September 2000.

48. E. J. Carl, A. B. John, and G. T. Neil, "Improving plant reliability through corrosion monitoring," inProceedings of the Process Plant Reliability, Houston, Tex, USA, November 1995.

Citations

CHAPTER 1

Matjaž Finšgar and Jennifer Jackson, Application of corrosion inhibitors for steels in acidic media for the oil and gas industry: A review, doi:10.1016/j.corsci.2014.04.044.

CHAPTER 2

M. Esmaily, M. Shahabi-Navid, J.-E. Svensson, M. Halvarsson, L. Nyborg, Y. Cao, and L.-G. Johansson, Influence of Temperature on the Atmospheric Corrosion of the Mg–Al Alloy Am50, doi:10.1016/j.corsci.2014.10.040.

CHAPTER 3

Xian Zhang, Inger Odnevall Wallinder, Christofer Leygraf, Mechanistic studies of corrosion product flaking on copper and copper-based alloys in marine environments, doi:10.1016/j.corsci.2014.03.028.

CHAPTER 4

Osokogwu, U and Oghenekaro .E, Evaluation of Corrosion Inhibitors Effectiveness in Oilfield Production Operations, ISSN 2277-8616.

CHAPTER 5

Mahmoud Abbas Ibraheem, Abd El Aziz El Sayed Fouda, Mohamed Talaat Rashad, and Fawzy Nagy Sabbahy, "Sweet Corrosion Inhibition on API 5L-B Pipeline Steel," ISRN Metallurgy, vol. 2012, Article ID 892385, 15 pages, 2012. doi:10.5402/2012/892385.

CHAPTER 6

B. Valdez, M. Schorr, R. Zlatev, M. Carrillo, M. Stoytcheva, L. Alvarez, A. Eliezer, and N. Rosas (2012). Corrosion Control in Industry, Environmental and Industrial Corrosion - Practical and Theoretical Aspects, Dr. Benjamin Valdez (Ed.), ISBN: 978-953-51-0877-1, InTech, DOI: 10.5772/51987.

CHAPTER 7

Chinedu I. Ossai, "Advances in Asset Management Techniques: An Overview of Corrosion Mechanisms and Mitigation Strategies for Oil and Gas Pipelines," ISRN Corrosion, vol. 2012, Article ID 570143, 10 pages, 2012. doi:10.5402/2012/570143

Index